RT-Thread
实时操作系统
内核、驱动和应用开发技术

郑苗秀　　沈鸿飞　　廖建尚 ◎ 编著

电子工业出版社
Publishing House of Electronics Industry
北京·BEIJING

内 容 简 介

RT-Thread 是我国自主研发的一款嵌入式实时多线程操作系统，专门设计用于嵌入式系统和物联网设备。本书主要介绍 RT-Thread 开发技术，由浅入深地介绍了 RT-Thread 的基础知识、开发环境与工具、内核开发技术、设备驱动开发技术、文件系统开发技术、GUI 开发技术和网络开发技术。本书边介绍理论知识边介绍开发技术，将理论学习和开发实践紧密结合起来，并给出了相关案例的完整代码，读者可以在案例代码的基础上快速地进行二次开发。

本书既可作为高等院校相关专业的教材或教学参考书，也可供相关领域的工程技术人员查阅。对于嵌入式、物联网和智能硬件开发的爱好者来说，本书也是一本深入浅出、贴近实际应用的技术读物。

本书提供了详尽的工程应用技术资料、案例代码和配套 PPT，读者可登录华信教育资源网（www.hxedu.com.cn）免费注册后下载。

图书在版编目（CIP）数据

RT-Thread 实时操作系统内核、驱动和应用开发技术 / 郑苗秀，沈鸿飞，廖建尚编著. -- 北京 : 电子工业出版社，2024. 7. --（新工科人才培养系列丛书）. -- ISBN 978-7-121-48650-0

Ⅰ. TP316.2

中国国家版本馆 CIP 数据核字第 2024XP8800 号

责任编辑：田宏峰
印　　刷：固安县铭成印刷有限公司
装　　订：固安县铭成印刷有限公司
出版发行：电子工业出版社
　　　　　北京市海淀区万寿路 173 信箱　邮编 100036
开　　本：787×1 092　1/16　印张：20.25　字数：518 千字
版　　次：2024 年 7 月第 1 版
印　　次：2025 年 4 月第 2 次印刷
定　　价：88.00 元

凡所购买电子工业出版社图书有缺损问题，请向购买书店调换。若书店售缺，请与本社发行部联系，联系及邮购电话：（010）88254888，88258888。

质量投诉请发邮件至 zlts@phei.com.cn，盗版侵权举报请发邮件至 dbqq@phei.com.cn。

本书咨询联系方式：tianhf@phei.com.cn。

党的二十大报告指出："以国家战略需求为导向，集聚力量进行原创性引领性科技攻关，坚决打赢关键核心技术攻坚战。"党的二十届三中全会提出："坚持面向世界科技前沿、面向经济主战场、面向国家重大需求、面向人民生命健康，优化重大科技创新组织机制，统筹强化关键核心技术攻关，推动科技创新力量、要素配置、人才队伍体系化、建制化、协同化。"

操作系统是实现网络强国的关键基石，对于提升国家网络空间竞争力、实现高水平科技自立自强具有重要意义，必须加快操作系统国产替代研发。

RT-Thread 诞生于 2006 年，是国内以开源中立、社区化发展起来的一款完全自主的实时操作系统。经过多年的发展，RT-Thread 以其高可靠性、高安全性、高可伸缩性和丰富的中间件，极大地满足了市场需求，目前已经成为市面上装机量最大（超 20 亿台）、开发者数量最多（超 20 万人）、软/硬件生态最好的操作系统之一，被广泛应用于航空、电力、轨道交通、汽车、工业自动化、消费电子等领域。

RT-Thread 内置了智能 AI 引擎，集成了与音频、图像相关的各类算法和智能引擎，也是 AIoT 领域的主流操作系统之一。

本书介绍 RT-Thread 开发技术，主要内容如下：

第 1 章为 RT-Thread 概述与开发基础。本章首先介绍了 RT-Thread 的特点、发展过程、原理架构和应用领域，然后演示了 RT-Thread 开发套件和 RT-Thread Studio 的开发方法，接着分析了 RT-Thread 移植技术，最后介绍了 FinSH 控制台的应用。

第 2 章为 RT-Thread 内核开发技术。本章主要介绍了线程管理、定时器、信号量、互斥量、事件集、邮箱、消息队列、信号和内存管理等的基本概念与工作原理，并以此为基础介绍内核开发技术。

第 3 章为 RT-Thread 设备驱动开发技术。本章主要介绍了 IO 设备、UART 设备、PIN 设备、ADC 设备、HWTIMER 设备、I2C 设备、PWM 设备、RTC 设备、SPI 设备、WATCHDOG 设备和 SENSOR 设备的基本概念与工作原理，在此基础上介绍了 RT-Thread 的管理方式，并介绍了相应接口的应用开发。

第 4 章为 RT-Thread 文件系统开发技术。本章主要介绍了文件系统中的挂载管理、文件管理和目录管理的基本概念与工作原理，在此基础上介绍了 RT-Thread 的管理方式，并介绍了相应接口的应用开发。

第 5 章为 RT-Thread GUI 开发技术。本章主要介绍了 GUI 基础和 emWin 图形库、GUI 图形和颜色、GUI 文本显示、GUI 图像显示、GUI 控件的基本概念与工作原理，在此基础上介绍了 RT-Thread 的管理方式，并介绍了相应接口的应用开发。

第 6 章为 RT-Thread 网络应用开发技术。本章主要介绍了 LWIP、AT Socket 协议栈、

MQTT 协议和 HTTP 的基本概念与工作原理，在此基础上介绍了 RT-Thread 的管理方式，并介绍了相应接口的应用开发。

本书将 RT-Thread 的理论知识和实际案例结合起来，边介绍理论知识边进行开发，有助于读者快速掌握理论知识和开发技巧。针对每个实际案例，本书均提供了完整的代码，读者可以在案例代码的基础上快速地进行二次开发，方便地将这些案例转化为各种比赛和创新创业的案例。本书的案例也为工程技术开发人员和科研人员的工程设计及科研提供了较好的参考资料。

本书既可作为高等院校相关专业的教材或教学参考书，也可供相关领域的工程技术人员查阅。对于嵌入式、物联网和智能硬件开发的爱好者来说，本书也是一本深入浅出、贴近实际应用的技术读物。

感谢上海睿赛德电子科技有限公司和中智讯（武汉）科技有限公司在本书编写过程中提供的帮助！感谢电子工业出版社在本书出版过程中给予的大力支持！

在本书编写过程中，我们借鉴和参考了国内外专家、学者、技术人员的相关研究成果，我们尽可能按学术规范予以说明，但难免会有疏漏之处，在此谨向有关作者表示深深的敬意和谢意，如有疏漏，请及时通过出版社与我们联系。

由于本书涉及的知识面广，编写时间仓促，以及我们的水平和经验有限，本书中的疏漏之处在所难免，恳请广大专家和读者批评指正。

作　者

2024 年 7 月

目　录

第 1 章
RT-Thread 概述与开发基础

1.1 RT-Thread 概述、优点与应用领域

1.1.1 RT-Thread 概述

RT-Thread 的全称是 Real Time-Thread，诞生于 2006 年，是国内开源中立的、以社区化形式发展起来的一款高可靠实时操作系统，由上海睿赛德电子科技有限公司（简称睿赛德）负责开发维护和运营。基于十几年的沉淀积累，专业化的运营推广，丰富易用的中间件，以及高可靠性、高安全性、高可伸缩性等特性，RT-Thread 极大地满足了市场需求，目前已成为市面上装机量最大（超 20 亿台）、开发者最多（超 20 万人）、软/硬件生态最好的操作系统之一，被广泛应用于航空、电力、轨道交通、汽车、工业自动化、消费电子等众多行业。RT-Thread 的标识如图 1.1 所示。

图 1.1   RT-Thread 的标识

RT-Thread 不仅是实时性操作系统，也是一个完备的应用系统，包含了嵌入式系统所需的各种组件，如 TCP/IP 协议栈、文件系统、C 库接口和 GUI 等，具有体积小、启动快速、占用资源少、实时性高的特点。

RT-Thread 的底层主要采用 C 语言编写，性能优秀，可适配市面上众多的开发板，移植方便。对于资源受限的微处理器，可使用平台工具裁减出仅需 3 KB Flash、1.2 KB RAM 的 Nano 版本；对于资源丰富的设备，可以使用在线软件包配置工具快速地进行模块化，通过导入软件安装包来实现语音交互、多媒体、滑动控制等复杂的功能。

RT-Thread 的文档资料丰富，其官方网站的文档中心包含系统简介、内核、设备与驱动、组件及开发工具等内容，每部分内容都包含说明和案例，并提供详细的 API 参考手册，可供开发者学习和参考。

RT-Thread 的发展历程如图 1.2 所示。RT-Thread 的项目始于 2006 年，最初是由中国开发者 Bernie 开发的一个小型嵌入式实时操作系统。RT-Thread 经历了从开始的小型项目到一个功能丰富、灵活且备受欢迎的嵌入式系统的演变，它的发展离不开全球嵌入式开发者社区的积极参与和推动。

图 1.2　RT-Thread 的发展历程

1.1.2　RT-Thread 的优点

RT-Thread 的优点如下：

（1）微内核架构。RT-Thread 采用微内核架构，将系统内核分为核心部分和可选部分。核心部分包括线程调度器、任务管理和基础服务，其他功能（如文件系统、网络协议栈等）属于可选部分，是以组件的形式存在的。这种架构设计使 RT-Thread 更加轻量化、模块化，用户可以根据应用需求选择不同的功能模块。

（2）多线程。RT-Thread 支持多线程操作，允许在一个应用程序中同时运行多个任务，每个线程都有自己独立的内存堆和上下文，有助于提高系统的并发性和灵活性。

（3）实时性。RT-Thread 注重实时性能，通过实时线程调度器实现了对任务的高效调度。实时性能对嵌入式系统来说是至关重要的，尤其是在对响应时间有严格要求的应用中，如工业控制和汽车电子系统。

（4）设备驱动框架。RT-Thread 提供了设备驱动框架，允许用户方便地添加和管理硬件设备的驱动程序，有助于提高系统对外部硬件的兼容性，同时使得硬件抽象变得更容易。

（5）软件包管理系统。RT-Thread 引入了软件包管理系统，使用户可以方便地选择和管理他们所需的软件包，常用的软件包有各种驱动、文件系统、网络协议栈等，有助于简化系统的配置和定制。

（6）中断服务机制。嵌入式系统通常需要处理中断，RT-Thread 提供了中断服务机制，可用于实时响应外部事件，这对外设通信、传感器输入等操作而言是至关重要的。

（7）支持多种处理器架构。RT-Thread 可以移植到多种处理器架构上，包括 ARM、MIPS、PowerPC 等，适用于不同类型的硬件平台。

（8）线程调度器。RT-Thread 的线程调度器负责任务的调度和管理，确保任务按照其优先级和时间片轮询的方式执行，有助于实现任务的合理分配和实时性能的优化。

（9）内存管理。RT-Thread 提供了内存管理机制，包括动态内存的分配和释放，这对嵌入式系统来说很重要，因为嵌入式系统的资源通常是受限的。

（10）社区支持和开源。RT-Thread 是一个开源项目，拥有庞大的开发者社区，社区的支持和贡献推动着 RT-Thread 不断改进和更新。

RT-Thread 以其微内核架构、实时性能和灵活性的特点，为嵌入式系统提供了一款功能强大的实时操作系统，其开源性质和活跃的社区生态也使 RT-Thread 成为嵌入式开发者的首选。

1.1.3　RT-Thread 的应用领域

RT-Thread 在多个领域都有广泛的应用，以下是 RT-Thread 的常见应用领域：

（1）工业自动化。RT-Thread 在工业控制系统中被广泛用于实时检测和控制，RT-Thread 的实时性和多线程使其非常适合处理工业自动化中的实时任务，如运动控制和传感器的数据采集。

（2）医疗设备。在医疗设备领域，RT-Thread 常用来检测和控制各种医疗设备，如呼吸机、心电图仪、血压计等。

（3）物联网（IoT）设备。RT-Thread 适用于物联网设备，如智能家居系统、智能传感器和嵌入式物联网网关，RT-Thread 的轻量级设计使其能够在资源受限的设备上运行。

（4）汽车电子系统。在汽车领域，RT-Thread 可以用于车载控制系统、嵌入式信息娱乐系统、辅助驾驶系统等。

（5）家电和消费电子。RT-Thread 可用来控制家用电器，如智能电视、智能空调、智能冰箱等，RT-Thread 的多线程和可移植性使其适用于不同的嵌入式设备。

（6）网络设备。RT-Thread 可以用于网络设备，如路由器、交换机和网络摄像头，RT-Thread 的网络协议栈和设备驱动框架使其适合处理网络通信任务。

（7）军事和航空航天。军事和航空航天领域对实时性和可靠性的要求非常高，RT-Thread 可用于飞行控制系统、导弹控制系统等。

（8）能源管理。在能源领域，RT-Thread 可以用于检测和控制能源设备，如太阳能发电控制系统、风力发电控制系统等。

（9）教育和研究。RT-Thread 也广泛应用于教育和研究领域，研究人员可通过 RT-Thread 来学习和研究嵌入式实时系统。

RT-Thread 在多个应用领域都表现出很强的灵活性，这使其成为嵌入式系统项目的首选。RT-Thread 的开源性和丰富的社区支持为开发者提供了丰富的资源和工具。

1.2 RT-Thread 的开发基础

近年来，物联网（Internet of Things，IoT）的应用得到普及，物联网市场发展迅猛，嵌入式设备的联网已成大势所趋。终端联网使软件复杂性大幅增加，传统的 RTOS 内核越来越难以满足市场需求。在这种情况下，物联网操作系统（IoT OS）的概念应运而生。物联网操作系统是指以操作系统内核（可以是 RTOS、Linux 等）为基础，包括诸如文件系统、图形库等较为完整的中间件，具备低功耗、安全、通信协议栈和云端连接能力的软件平台。RT-Thread 就是一个 IoT OS。

本书实例的开发环境为：PC 的处理器为 2 GHz 的 Pentium 双核处理器、内存大小为 4 GB、操作系统为 64 位的 Windows7 及以上版本，开发软件为 RT-Thread Studio 开发软件，硬件开发平台为嵌入式原型机 ZI-ARMEmbed。

1.2.1 原理分析

本节的要求如下：

➲ 学习 RT-Thread 的基础知识。

➲ 了解 RT-Thread 的架构。

➲ 掌握 RT-Thread 的基本开发方法。

1.2.1.1 RT-Thread 的架构

RT-Thread 与其他 RTOS（如 FreeRTOS、μC/OS）的主要区别之一是，它不仅是一个实时的嵌入式操作系统，还具备丰富的中间件，其架构如图 1.3 所示。

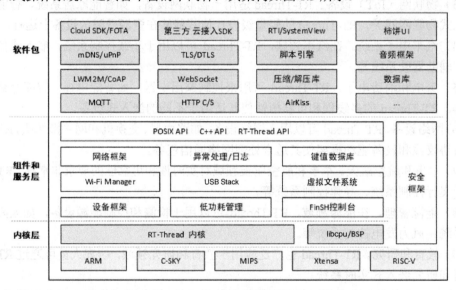

图 1.3   RT-Thread 的架构

RT-Thread 主要包括以下部分：

（1）内核层：RT-Thread 内核是内核层的核心部分，包括内核系统中对象的实现，如多线程及其调度、信号量、邮箱、消息队列、内存管理、定时器等；libcpu/BSP（芯片移植相关文件/板级支持包）与硬件密切相关，由外设驱动程序和 CPU 移植代码构成。

（2）组件和服务层：组件是基于 RT-Thread 内核之上的软件，如虚拟文件系统（DFS）、FinSH 控制台、网络框架、设备框架等。RT-Thread 采用模块化设计，实现了组件内部的高内聚、组件之间的低耦合。

（3）软件包：运行在 RT-Thread 内核之上，主要包括面向不同应用领域的通用软件（通常由描述信息、源代码或库文件组成）。RT-Thread 提供了开放的软件包，软件包中存放的是 RT-Thread 官方或开发者提供的软件，为用户提供了众多可重用的软件，是 RT-Thread 生态的重要组成部分。

1.2.1.2 RT-Thread 的硬件开发平台

1）ZI-ARMEmbed

本书使用的嵌入式原型机是 ZI-ARMEmbed。ZI-ARMEmbed 是中智讯（武汉）科技有限公司推出的 RT-Thread 原型机，获得了 RT-Thread 官方（上海睿赛德电子科技有限公司）的认证。ZI-ARMEmbed 实物如图 1.4 所示，该原型机采用的是 STM32F407 微控制器（基于 ARM Cortex-M4 内核）。

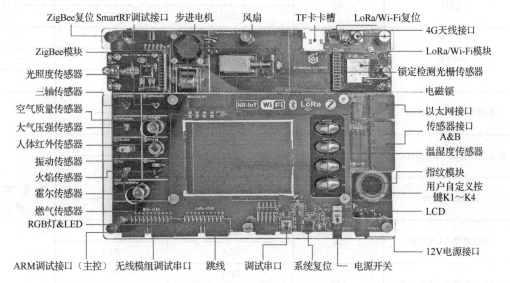

图 1.4　ZI-ARMEmbed 实物

ZI-ARMEmbed 的部分模块或传感器如表 1.1 所示。

表 1.1　ZI-ARMEmbed 的部分模块或传感器

部分模块或传感器	通信方式	引脚名称或引脚数量	对应的微控制器引脚	备　注
指纹模块	UART	UART_RX	PC6	串口 6
		UART_TX	PC7	
	IO	TOUCH_OUT	PC5	—
空气质量传感器（MP503）	ADC	4	PC3（模拟量）	—
人体红外传感器（AS312）	IO	2	PC8	—
电机（A3967SLB）	IO	ENABLE	PC9	—
		DIR	PE4	—
		STEP	PB6	—
风扇	IO	—	PE5	—
火焰传感器	IO	—	PE6	—
霍尔传感器	IO	3	PD11	—
锁定检测光栅传感器	IO	—	PD13	—
燃气传感器（MP-4）	ADC	4	PC4（模拟量）	—
振动传感器	IO	—	PA12	—
电磁锁	IO	—	PA11	—
LED	IO	LED1	PE0	—
		LED2	PE1	—

续表

部分模块或传感器	通信方式	引脚名称或引脚数量	对应的微控制器引脚	备　　注
LED	IO	LED3	PE2	—
		LED4	PE3	—
RGB 灯	IO	R	PB0	—
		G	PB1	—
		B	PB2	—
光照度传感器（BH1750）	I2C	SCL	PA1	同一组 I2C 总线
		SDA	PA0	
温湿度传感器（HTU21D）	I2C	SCL	PA1	
		SDA	PA0	
PCF8563 时钟	I2C	SCL	PA1	
		SDA	PA0	
三轴传感器（LIS3DH）	I2C	SCL	PA1	
		SDA	PA0	
大气压强传感器（FBM320）	I2C	SDA	PA1	
		SCL	PA0	
蜂鸣器	IO	BEEP	PD6	—
RELAY-1	IO	RELAY-1	PC12	—
RELAY-2	IO	RELAY-2	PC13	—
按键	IO	KEY1	PB12	—
		KEY2	PB13	—
		KEY3	PB14	—
		KEY4	PB15	—

2）STM32 系列微控制器

ZI-ARMEmbed 采用的微控制器是 STM32F407，属于 STM32 系列微控制器。STM32 系列微控制器的型号、内核和特点如表 1.2 所示。

表 1.2　STM32 系列微控制器的型号、内核和特点

型　　号	内　　核	特　　点
STM32F0	Cortex-M0	低成本，属于入门级的微控制器
STM32L0	Cortex-M0+	低功耗
STM32F1	Cortex-M3	通用型微控制器
STM32F2		存储空间大，采用硬件加密
STM32L1		低功耗
STM32M		集成了遵循 IEEE 802.15.4 标准的无线通信模块
STM32F3	Cortex-M4	模拟通道，具有更灵活的数据通信矩阵
STM32F4		在 168 MHz 的时钟频率下，可以零等待地访问 Flash；具有动态功率调整技术
STM32F7	Cortex-M7	L1 一级缓存，时钟频率高达 200 MHz

1.2.1.3 ZI-ARMEmbed 的连接

1）ARM 仿真器的连接

通过 ARM 调试接口可以连接 ARM 仿真器与 ZI-ARMEmbed，如图 1.5 所示。

2）串口的连接

（1）使用 MiniUSB 线可以连接 ZI-ARMEmbed 与 PC。串口连接如图 1.6 所示。

图 1.5　ARM 仿真器与 ZI-ARMEmbed 的连接　　　　图 1.6　串口连接

（2）右键单击"我的电脑"，在弹出的右键菜单中选择"管理"可进入"计算机管理"界面。

（3）单击"设备管理器"，找到"端口（COM 和 LPT）"，查看串口的端口号。例如，COM5（每台计算机的端口号可能不一样）。若没有找到端口则有两种可能：第一种是没有按照实验要求使用 MiniUSB 线连接硬件与 PC（重新检查连接即可）；另一种是没有安装串口驱动（串口在第一次使用时需要安装驱动程序，只需要根据相应提示即可安装驱动程序）。串口驱动程序分为 CP210x_Drivers.zip 或者 CH340_Drivers.zip，用户可根据实际的硬件来选择不同的驱动程序。

情况 1：安装 CP210x_Drivers.zip 的结果如图 1.7 所示。

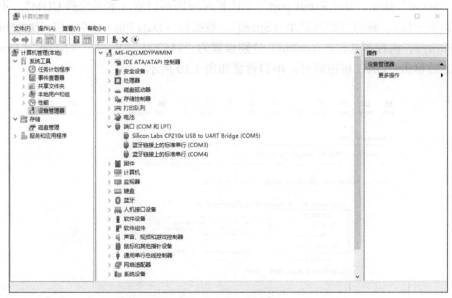

图 1.7　安装 CP210x_Drivers.zip 的结果

情况 2：安装 CH340_Drivers.zip 的结果如图 1.8 所示。

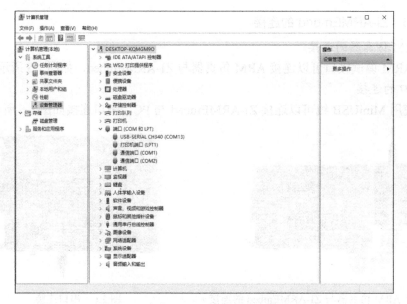

图 1.8　安装 CH340_Drivers.zip 的结果

1.2.1.4　串口工具 MobaXterm

1）串口设置

（1）安装 MobaXterm。双击 MobaXterm.exe 文件（见图 1.9）即可安装 MobaXterm。

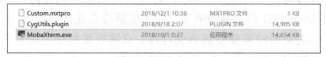

图 1.9　MobaXterm.exe 文件

（2）启动 MobaXterm 后单击 "Session" 可打开 "Session settings" 对话框，在该对话框中单击 "Serial" 按钮，在 "Serial port" 中选择对应的端口号（本书选择 COM5，不同的计算机可能不一样）。将串口的波特率（Speed）、数据位（Data bits）、停止位（Stop bits）、校验位（Parity）、流控制（Flow control）分别设置为 "115200" "8" "1" "None" "None"，设置完成之后单击 "OK" 按钮即可。串口设置如图 1.10 所示。

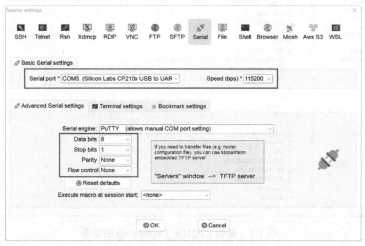

图 1.10　串口设置

2）串口使用

（1）为 ZI-ARMEmbed 上电。

（2）进入 FinSH 控制台（见图 1.11）后，在 MobaXterm 的串口终端可以看到启动消息。

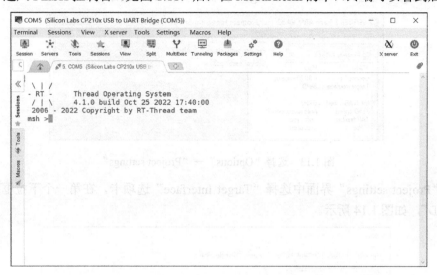

图 1.11　FinSH 控制台

1.2.1.5　烧写工具 J-Flash ARM

1）J-Flash ARM 的安装

烧写工具 J-Flash ARM 可将 hex 文件烧写到 ARM 内核的单片机（如 STM32F407）中。双击安装包文件，即可默认安装该工具。特别说明：在安装 J-Flash ARM 时，如果已经安装了 IAR for ARM 等其他烧写工具，会弹出如图 1.12 所示的提示框。

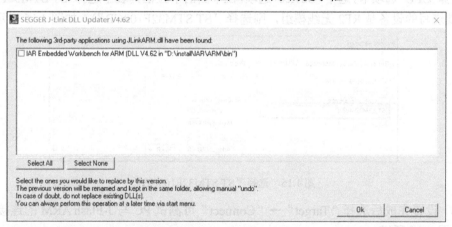

图 1.12　安装了其他烧写工具的提示框

该提示框提示用户是否换掉 IAR for ARM 安装路径下的 JLinkARM.dll 文件。单击"Select All"按钮，勾选"IAR Embedded Workbench for ARM（DLL V4.62 in "D:\install\IAR\ARM\bin"）"选项后，单击"OK"按钮即可按照安装向导来完成 J-Flash ARM 的安装。

2）J-Flash ARM 的设置

（1）安装 J-Flash ARM 后，单击开始菜单栏中"SEGGER"下的 J-Flash ARM 软件，即

可启动 J-Flash ARM。

（2）在 J-Flash ARM 界面中，用户可根据实际的硬件需要进行设置。选择"Options"→"Project settings"可进入"Project settings"界面，如图 1.13 所示。

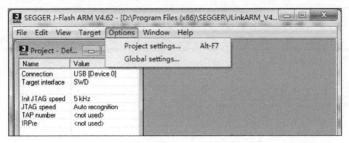

图 1.13　选择"Options"→"Project settings"

在"Project settings"界面中选择"Target Interface"选项卡，在第一个下拉框列表中选择"SWD"，如图 1.14 所示。

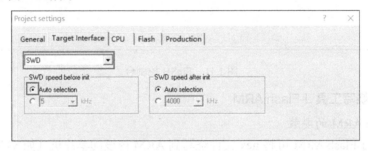

图 1.14　选择"SWD"

选择"CPU"选项卡，选中"Device"选项，在下面的下拉框列表中选择"ST STM32F407VE"（如果要烧写的设备是 RF2 无线模组，则选择"ST STM32F103CB"），如图 1.15 所示。

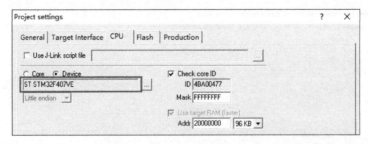

图 1.15　选择"ST STM32F103CB"

（3）设置完成后选择"Target"→"Connect"可测试 PC 与 J-Flash ARM 的连接是否正常，如图 1.16 所示。

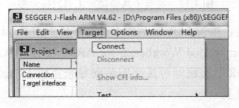

图 1.16　选择"Target"→"Connect"

连接成功后，J-Flash ARM 界面的下方会提示"Connected successfully"，如图 1.17 所示。

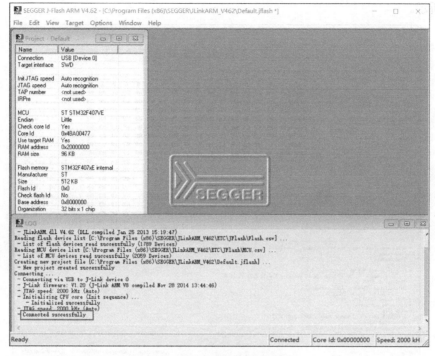

图 1.17　连接成功的提示

3）通过 J-Flash ARM 烧写固件（hex 文件）

（1）在 J-Flash ARM 界面中选择"File"→"Open data file"，打开要烧写到 ZI-ARMEmbed 的 hex 文件，如图 1.18 所示。

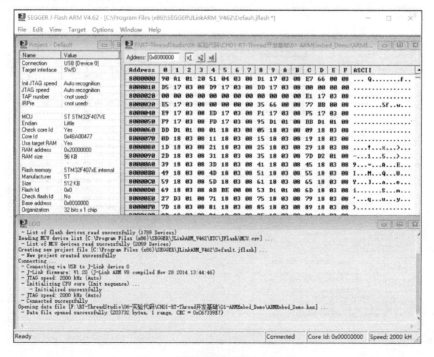

图 1.18　选择要烧写的 hex 文件

（2）选择"Target"→"Erase chip"，擦除 MCU 的 Flash 扇区，擦除完成后的提示如图 1.19 所示。

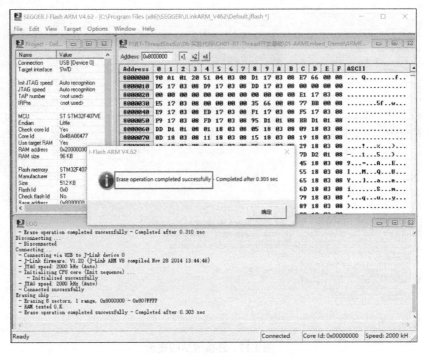

图 1.19　擦除完成后的提示

（3）选择"Target"→"Program & Verify"，将选中的 hex 文件烧写到 ZI-ARMEmbed 的 MCU 中，烧写完成后的提示如图 1.20 所示。

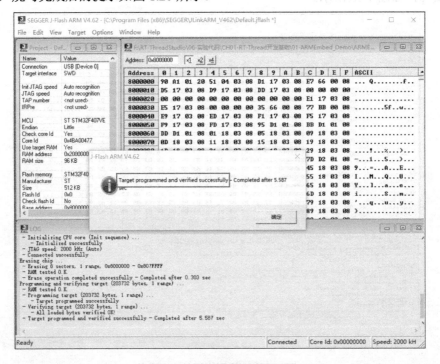

图 1.20　烧写完成后的提示

1.2.2　开发设计与实践

本项目使用 J-Flash ARM 烧写 hex 文件，使用 ZI-ARMEmbed 的 LCD 显示人机交互界面，K1～K4 四个功能按键用来和 ZI-ARMEmbed 进行交互。

1.2.3　开发步骤与验证

1.2.3.1　硬件部署

（1）准备 ZI-ARMEmbed、ARM 仿真器、MiniUSB 线。

（2）完成仿真器、串口和 12 V 电源之间的连接。

（3）使用 MobaXterm 工具创建串口终端，将串口的波特率、数据位、停止位、校验位、流控制分别设置为 115200、8、1、None、None。

1.2.3.2　程序下载

（1）运行 J-Flash ARM，选择要烧写的 hex 文件。

（2）选择 J-Flash ARM 界面中的"Target →Program &Verify"，将 hex 文件烧写到 ZI-ARMEmbed 的 STM32F407 中。

（3）拔掉 ARM 仿真器，按下复位键启动 ZI-ARMEmbed。

1.2.3.3　验证效果

1）ZI-ARMEmbed

程序下载成功之后，ZI-ARMEmbed 的 LCD 点亮，并显示传感器状态信息。用户可以通过 K1～K4 四个功能按键与 ZI-ARMEmbed 进行交互，K1 用于翻页，K2 用于选择和取消选择，K3 和 K4 用于上、下移动。ZI-ARMEmbed 的 LCD 显示的页面如图 1.21 所示。

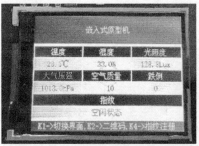

（a）显示各种采集类传感器数据的页面

（b）显示各种控制类传感器状态及开关操作的页面

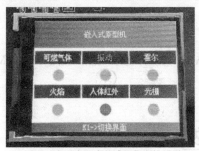

（c）显示各种安防类传感器状态的页面

（d）显示无线设备参数的页面

图 1.21　ZI-ARMEmbed 的 LCD 显示的页面

2）FinSH 控制台

（1）启动 MobaXterm 后，可在串口终端看到启动消息，进入 FinSH 控制台。

（2）在 MobaXterm 串口终端的 FinSH 控制台输入"help"命令（输入之前需要把输入法切换到英文）后按 Enter 键，可输出 ZI-ARMEmbed 的命令，通过这些命令可得到不同的输出信息。

```
    \ | /
  - RT -      Thread Operating System
   / | \        4.0.2 build Apr 27 2022
  2006 - 2020 Copyright by rt-thread team
msh >finger_request_with_data 35
finger mode info: ID_SEODU2_ID812_BYD160_Inne
[RF2 <-->   LoRa ]
[RF1 <--> ZigBee]
msh >help
RT-Thread shell commands:
clear              - clear the terminal screen
version              - show RT-Thread version information
list_thread        - list thread
list_sem            - list semaphore in system
list_event          - list event in system
list_mutex          - list mutex in system
list_mailbox        - list mail box in system
list_msgqueue        - list message queue in system
list_mempool        - list memory pool in system
list_timer          - list timer in system
list_device          - list device in system
gateway              - eth gateway set
zigbee              - zigbee terminal set
lora                - lora terminal set
nb                  - nb terminal set
wifi                - wifi terminal set
ble                - ble terminal info
reboot              - System reboot.
help                - RT-Thread shell help.
ps                  - List threads in the system.
free                - Show the memory usage in the system.
```

1.2.4　小结

本节首先介绍了 RT-Thread 的架构、J-Flash ARM 烧写工具，然后使用 ZI-ARMEmbed 的 LCD 显示人机交互页面。

1.3 RT-Thread Studio 的应用开发

集成开发环境（Integrated Development Environment，IDE）是一种用于嵌入式系统开发

的集成工具，提供了丰富的工具和资源，可帮助开发人员创建、编译、调试和测试嵌入式系统的软件和硬件。以下是一些常用的 IDE：

（1）Keil μVision：是由 Keil 公司（一家知名的嵌入式开发工具提供商）开发的，广泛用于 ARM Cortex-M 微控制器和其他嵌入式系统的开发，包括编译器、调试器、仿真器和各种插件。

（2）IAR Embedded Workbench：IAR Systems 公司的 IAR Embedded Workbench 是另一个备受欢迎的 IDE，可用于多种嵌入式平台，包括 ARM、MSP430 和 Renesas 等。

（3）Atmel Studio：是由 Microchip Technology 公司（之前是 Atmel Corporation）为其 AVR 和 SAM 微控制器提供的 IDE，集成了编译器、调试工具和 Atmel START（用于快速启动项目的库）。

（4）Eclipse：Eclipse 是一个免费的、开源的 IDE，可通过扩展各种插件来开发嵌入式系统。Eclipse 通常需要额外的插件来支持特定的硬件架构和编程语言。

（5）PlatformIO：PlatformIO 是一个用于嵌入式开发的开源 IDE，支持多种平台和硬件架构，包括 Arduino、ESP8266、ESP32、STM32 等。

（6）Code Composer Studio：Code Composer Studio 是德州仪器（Texas Instruments）为其 Tiva C 和 CCS 系列微控制器提供的集成开发环境。

（7）RT-Thread Studio：是一个基于 RT-Thread 的 IDE，专门用于设计嵌入式系统和物联网系统。

本节的要求如下：

⊃ 了解 RT-Thread Studio 软件。

⊃ 掌握 RT-Thread Studio 新建工程的方法。

⊃ 掌握 RT-Thread Studio 编译、调试及下载程序等基本开发方法。

1.3.1　RT-Thread Studio 分析

RT-Thread Studio 专门用于开发基于 RT-Thread 的嵌入式系统和物联网系统，提供了一套工具，可帮助开发人员更容易地创建、调试和部署嵌入式系统。RT-Thread Studio 的主要特点和功能如下：

（1）以 Eclipse 为基础。RT-Thread Studio 的基础是 Eclipse（一个流行的开源综合性 IDE），这使得 RT-Thread Studio 可以利用 Eclipse 生态系统中的丰富插件和工具。

（2）支持多种处理器架构。RT-Thread Studio 支持多种处理器架构，包括 ARM、MIPS、RISC-V 等，这使得它适用于不同类型的嵌入式系统。

（3）集成了 RT-Thread。RT-Thread Studio 集成了 RT-Thread，提供了对 RT-Thread API 的支持，简化了 RT-Thread 项目的创建和配置。

（4）提供了图形化配置工具。RT-Thread Studio 提供了图形化配置工具，允许开发人员通过界面来配置 RT-Thread，无须手动编辑配置文件。

（5）支持多种调试工具。RT-Thread Studio 支持多种调试工具，包括 GDB 调试器，这使得开发人员能够进行全面的调试和性能分析。

（6）提供了项目管理工具。RT-Thread Studio 提供了项目管理工具，可方便开发人员管理复杂的嵌入式项目，包括代码的组织、构建和部署。

RT-Thread Studio 旨在简化嵌入式系统的开发流程，提高开发效率，特别是对于使用 RT-Thread 的项目。RT-Thread Studio 的运行界面如图 1.22 所示。

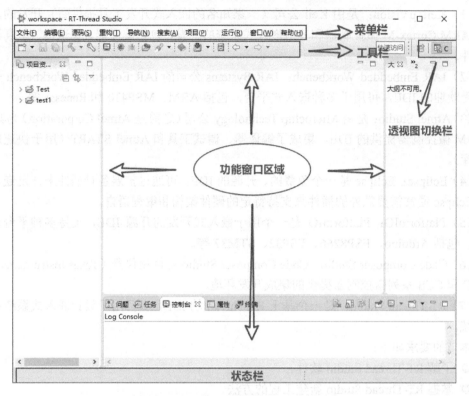

图 1.22 RT-Thread Studio 的运行界面

1.3.2 开发设计与实践

本节主要介绍 RT-Thread Studio 的安装和部署，使用 RT-Thread Studio 新建工程，编译和下载程序，帮助读者熟悉 RT-Thread Studio 的常用操作和程序调试。

1.3.3 开发步骤与验证

1.3.3.1 硬件部署

同 1.2.3.1 节。

1.3.3.2 安装并配置 RT-Thread Studio

（1）读者可以从本书的配套资源中找到 RT-Thread Studio 的安装包，也可以从 RT-Thread 官网下载 RT-Thread Studio 的安装包。本书建议读者使用配套资源中的安装包，该安装包已经集成了必备的系统内核和芯片 BSP。

（2）将本书配套资源中的 RT-ThreadStudio.zip（RT-Thread Studio 的安装包）解压缩（注意目录中不要有中文文字符），右键单击 studio.exe 文件，在弹出的右键菜单中选择"以管理员身份运行"，如图 1.23 所示，根据安装向导可以轻松完成 RT-Thread Studio 的安装，安装向导界面如图 1.24 所示。

图 1.23　选择"以管理员身份运行"

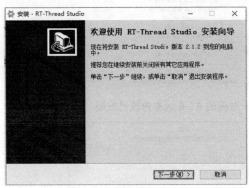

（a）开始安装

（b）接受许可协议

（c）选择目标位置

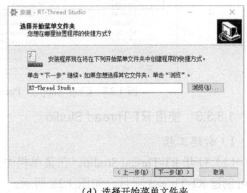

（d）选择开始菜单文件夹

（e）准备安装

（f）安装完成

图 1.24　安装向导界面

（3）安装完成后打开 RT-Thread Studio，单击工具栏（见图1.25）中的""（SDK Manager）按钮可以进行 SDK 软件包管理。

图 1.25 工具栏

（4）本书使用的 ZI-ARMEmbed 的芯片为 STM32F407，RT-Thread 的版本为 4.1.0，因此需要在 SDK 管理器（SDK Manager）中，确保 RT-Thread_Source_Code 对应的 4.1.0 版本内核和 Chip_Support_Packages 对应的 STM32F4 已安装，如图 1.26 和图 1.27 所示。

图 1.26 RT-Thread_Source_Code 对应的 4.1.0 版本内核已安装

图 1.27 Chip_Support_Packages 对应的 STM32F4 已安装

1.3.3.3 使用 RT-Thread Studio

1）新建工程

（1）打开 RT-Thread Studio，在菜单栏中选择"文件"→"新建"→"RT-Thread 项目"，即可创建 RT-Thread 项目，如图 1.28 所示。

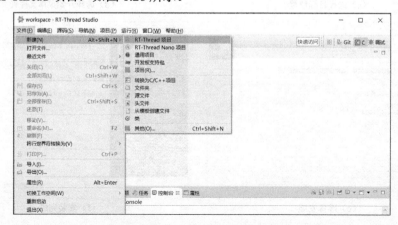

图 1.28 选择"文件"→"新建"→"RT-Thread 项目"

（2）根据图 1.29 对项目进行配置：将"Project name"（工程名）设为"02-Template"，将"RT-Thread"设为"4.1.0"，将"厂商"设为"STMicroelectronics"，将"系列"设为"STM32F4"，将"子系列"设为"STM32F407"，将"芯片"设为"STM32F407VE"，将"调试器"设为"J-Link"，将"接口"设为"SWD"，"控制台串口"、"发送脚"和"接收脚"需要根据具体的设备来设置，单击"完成"按钮即可创建"02-Template"工程。

图 1.29　项目配置界面

（3）创建好的"02-Template"工程如图 1.30 所示。

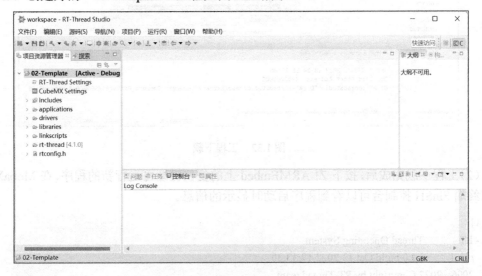

图 1.30　创建好的"02-Template"工程

2）工程编译

选中"02-Template"工程后，单击工具栏中的"🔧"（构建）按钮即可进行编译，当控制台窗口中提示编译完成并没有报错时，表示工程编译成功，如图1.31所示。

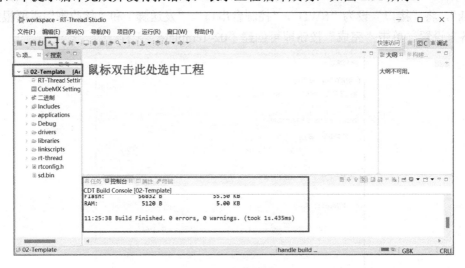

图 1.31 工程编译

3）工程下载

（1）单击工具栏中的"⬇"（下载）按钮可将编译后的工程下载到 ZI-ARMEmbed，当控制台窗口中提示执行完成且没有报错时，表示工程下载完成，如图1.32所示。

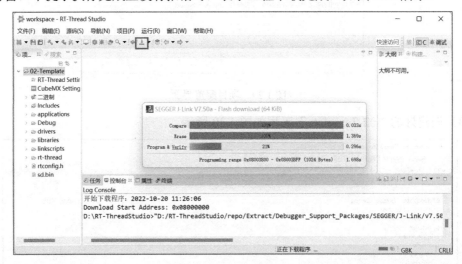

图 1.32 工程下载

（2）工程下载完成后，按下 ZI-ARMEmbed 上的复位键即可运行新的程序，在 MobaXterm 串口终端 FinSH 控制台可以看到程序启动时显示的信息。

```
       \ | /
     - RT -     Thread Operating System
      / | \     4.1.0 build Oct 13 2022 17:33:30
     2006 - 2022 Copyright by RT-Thread team
```

```
[D/main] Hello RT-Thread!
msh >[D/main] Hello RT-Thread!
[D/main] Hello RT-Thread!
```

4）工程调试

（1）单击工具栏中的"![bug]"（调试）按钮可启动调试功能，如图 1.33 所示。

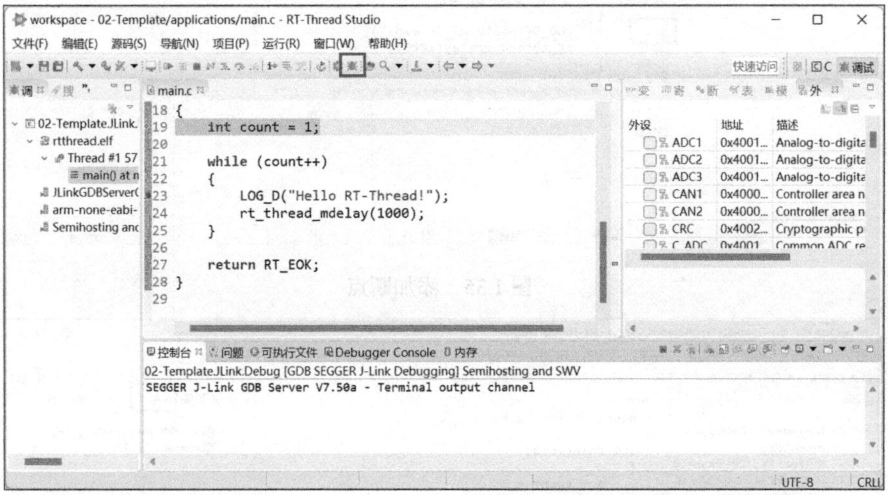

图 1.33　启动调试功能

（2）启动调试功能后，工具栏中会出现如图 1.34 所示的调试功能按钮，部分按钮的使用说明如表 1.3 所示。

图 1.34　工具栏中的调试功能按钮

表 1.3　部分调试功能按钮的使用说明

按 钮 图 标	使 用 说 明
![]	启动调试程序
![]	暂停调试程序，暂停后单击"![]"按钮后调试程序将继续运行
![]	停止调试，单击该按钮后将退出调试程序，即退出调试模式
![]	单步跳入，调试程序跳入，如跳入一个函数内部
![]	单步跳过，调试程序不会跳入具体的函数内部，可理解为单行调试
![]	单步返回，当进入函数内部时，单击该按钮可单步跳出该函数，回到外部函数
![]	命令步进模式，单击该按钮后，可实现汇编代码的单步调试

（3）设置和取消断点。在调试模式下，双击程序语句的序号即可添加断点，如图 1.35 所示。

单击工具栏中的"![]"按钮可全速运行程序，但当程序运行到断点处会停止。在右侧的断点管理器中可以查看所有的断点，也可以取消不需要的断点，如图 1.36 所示。

（4）跟踪变量。右侧的变量管理器（见图 1.37）会默认显示当前函数中的所有变量。

图 1.35　添加断点

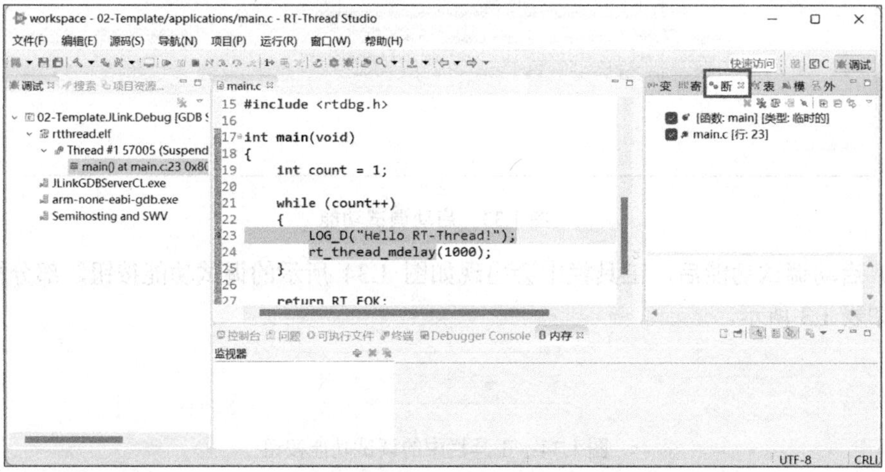

图 1.36　在断点管理器中查看断点

图 1.37　变量管理器

在表达式管理器（见图 1.38）中输入想要查看的变量名称后，即可显示该变量的相关信息。

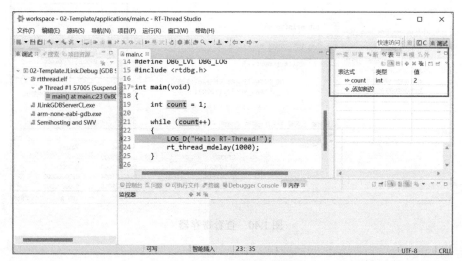

图 1.38 表达式管理器

（5）外设寄存器：由于 RT-Thread 的问题，在查看寄存器的数值时需要配置 CSP 的版本，在新建工程时选择的版本是 0.2.2，因此需要修改配置路径。单击工具栏中的"●"（调试配置）按钮可弹出"配置工程"对话框，在该对话框中单击"SVD path"选项卡，可将文件路径（file path）中的"0.2.0"修改为"0.2.2"，如图 1.39 所示。

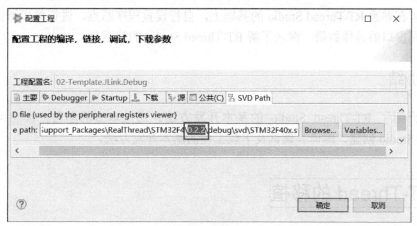

图 1.39 修改配置路径

修改完成后单击"确定"按钮，单击工具栏中的"▶"按钮即可正常显示寄存器的数值。单击相应的寄存器，就会在状态栏中显示具体的寄存器值，如图 1.40 所示。

（6）程序调试完后，单击工具栏中的"■"（停止）按钮可退出调试模式。

1.3.3.4 验证效果

（1）关闭 RT-Thread Studio，拔掉仿真器，按下 ZI-ARMEmbed 上的电源按键重新上电。

（2）本节创建的工程会默认地在 MobaXterm 串口终端 FinSH 控制台每秒输出一次"Hello RT-Thread!"。

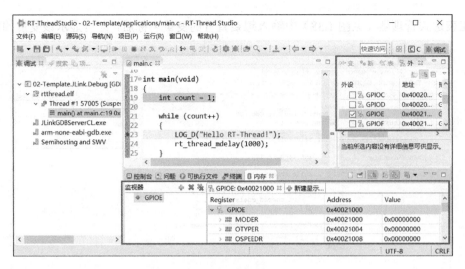

图 1.40　查看寄存器

```
        \ | /
      - RT -       Thread Operating System
       / | \       4.1.0 build Oct 13 2022 17:33:30
       2006 - 2022 Copyright by RT-Thread team
[D/main] Hello RT-Thread!
msh >[D/main] Hello RT-Thread!
[D/main] Hello RT-Thread!
```

读者可在熟悉 RT-Thread Studio 的基础上，自行设置程序断点，进行单步调试，并观察程序中不同窗口的具体数值，深入了解 RT-Thread Studio 的使用。

1.3.4　小结

本节介绍了 RT-Thread Studio 的基本开发方法。通过本节的学习，读者可掌握利用 RT-Thread Studio 创建、编译、调试及下载工程的基本开发方法。

1.4 RT-Thread 的移植

嵌入式系统的移植是指将嵌入式软件或操作系统从一个硬件平台移植到另一个硬件平台的过程。这通常需要适应目标硬件的架构、外设和资源，以确保嵌入式系统能够在新的目标系统上运行，该过程涉及处理器体系结构、外设、硬件抽象层、驱动程序和其他系统组件的变化。

本节的要求如下：

➲ 了解 RT-Thread 的启动流程。

➲ 学习 libcpu 和 BSP。

➲ 掌握 RT-Thread 移植的基本方法。

1.4.1　RT-Thread 的移植原理

1）RT-Thread 的启动流程

RT-Thread 的启动入口函数为 rtthread_startup()。当芯片的启动文件完成必要工作（如初始化时钟、配置中断向量表、初始化内存等）后，会跳转到 RT-Thread 的启动入口函数。RT-Thread 的启动流程如下：

（1）关闭全局中断，初始化与系统相关的硬件。

（2）显示系统版本信息，初始化系统内核对象（如定时器、线程调度器等）。

（3）初始化用户 main 线程（同时会初始化线程栈），在 main 线程中对各模块依次进行初始化。

（4）初始化软件定时器线程、初始化空闲线程。

（5）启动线程调度器，切换到第一个线程开始运行（如 main 线程），并打开全局中断。

2）启动文件（zonesion/common/LIB/Libraries/startup_stm32f407xx.S）

启动文件（startup_xx.S）由芯片厂商提供，位于芯片固件库中。每款芯片都有对应的启动文件，在不同开发环境下启动文件也不相同。当系统加入 RT-Thread 后，会将 RT-Thread 的启动放在调用 main() 函数之前，如图 1.41 所示。

图 1.41　RT-Thread 启动在调用 main() 之前

启动文件的主要工作包括初始化时钟、配置中断向量表、初始化全局/静态变量、初始化内存、初始化库函数、跳转程序等内容。

3）libcpu（rt-thread/libcpu）

RT-Thread 的 libcpu 向下提供了一套统一的 CPU 架构移植接口，这部分接口包含了全局中断开关函数、线程上下文切换函数、时钟节拍的配置和中断函数、Cache 函数等内容，RT-Thread 支持的 CPU 架构在源码的 libcpu 文件夹下。

4）上下文切换（rt-thread/libcpu/arm/cortex-m4/context_gcc.S）

上下文切换文件（context_xx.s）的主要工作是实现 CPU 在线程之间的切换或者线程与中断之间的切换等。在上下文切换中，CPU 一般会停止当前运行的代码，并保存当前程序运行的具体位置以便后续运行。上下文切换的接口如表 1.4 所示。

表 1.4　上下文切换的接口

需要实现的函数	描　述
rt_base_t rt_hw_interrupt_disable(void);	关闭全局中断
void rt_hw_interrupt_enable(rt_base_t level);	打开全局中断
void rt_hw_context_switch_to(rt_uint32 to);	在没有来源线程的上下文切换中，在第一个线程或者信号中启动线程调度器
void rt_hw_context_switch(rt_uint32 from, rt_uint32 to);	从线程 from 切换到线程 to，用于线程和线程之间的切换
Void rt_hw_context_switch_interrupt(rt_uint32 from, rt_uint32 to);	从线程 from 切换到线程 to，用于中断和线程之间的切换

5）线程栈初始化（rt-thread/libcpu/arm/cortex-m4/cpuport.c）

在 RT-Thread 中，线程具有独立的栈。在进行线程切换时，RT-Thread 会将当前线程的上下文保存在栈中；在线程要恢复运行时，再从栈中读取上下文信息。rt_hw_hard_fault_exception()为故障异常处理函数，在发生硬件错误时，会触发 HardFault_Handler 中断，从而调用该函数。

线程栈初始化文件（cpuport.c）的主要工作是实现线程栈的初始化和硬件错误的处理。线程栈初始化涉及的接口如表 1.5 所示。

表 1.5　线程栈初始化的接口

需要实现的函数	描　述
rt_hw_stack_init()	实现线程栈的初始化
rt_hw_hard_fault_exception()	硬件错误处理函数

6）中断与异常

在基于 Cortex-M 内核的微控制器中，中断是通过中断向量表来进行处理的。当某个中断被触发时，微控制器直接判定该中断属于哪个中断源，然后直接跳转到相应的固定位置进行处理，不需要再自行实现中断管理。

7）板级文件（zonesion/common/BSP/drivers/board.c 和 zonesion/common/BSP/drivers/board.h）

板级文件（board.c 和 board.h）的主要工作是实现 rt_hw_board_init()函数。该函数在 board.c 中，完成了系统启动必要的工作，主要包含：

（1）配置系统时钟。

（2）实现操作系统节拍。

（3）初始化外设（如 GPIO、UART 等）。

（4）设置控制台的串口设备。

（5）初始化系统内存，实现动态内存的管理。

（6）板级初始化，使用 INIT_BOARD_EXPORT()进行自动初始化的函数会在此处被初始化。

（7）其他必要的初始化。

板级初始化的代码如下：

```
/*******************************************************************************
 * 名称：rt_hw_board_init()
 * 功能：板级初始化
 *******************************************************************************/
RT_WEAK void rt_hw_board_init()
{
    extern void hw_board_init(char *clock_src, int32_t clock_src_freq, int32_t clock_target_freq);

    //Heap initialization
#if defined(RT_USING_HEAP)
    rt_system_heap_init((void *) HEAP_BEGIN, (void *) HEAP_END);   //内存初始化
#endif
```

```
    hw_board_init(BSP_CLOCK_SOURCE, BSP_CLOCK_SOURCE_FREQ_MHZ,
            BSP_CLOCK_SYSTEM_FREQ_MHZ);                    //板级初始化

    //Set the shell console output device
#if defined(RT_USING_DEVICE) && defined(RT_USING_CONSOLE)
    rt_console_set_device(RT_CONSOLE_DEVICE_NAME);          //设置控制台的串口设备名
#endif

    //Board underlying hardware initialization
#ifdef RT_USING_COMPONENTS_INIT
    rt_components_board_init();                             //初始化组件
#endif
}
```

在 rt_hw_board_init()函数中还会调用 hw_board_init()函数。hw_board_init()函数的主要工作是初始化 HAL 库、配置系统时钟、初始化 PIN 设备和串口设备等，代码如下：

```
/**********************************************************************************
 * 名称：hw_board_init()
 * 功能：初始化 HAL 库、配置系统时钟、初始化 PIN 设备和串口设备
 **********************************************************************************/
void hw_board_init(char *clock_src, int32_t clock_src_freq, int32_t clock_target_freq)
{
    extern void rt_hw_systick_init(void);
    extern void clk_init(char *clk_source, int source_freq, int target_freq);

#ifdef SCB_EnableICache
    /*Enable I-Cache--------------------------------------------------------*/
    SCB_EnableICache();
#endif

#ifdef SCB_EnableDCache
    /*Enable D-Cache--------------------------------------------------------*/
    SCB_EnableDCache();
#endif

    /*HAL_Init() function is called at the beginning of the program */
    HAL_Init();                                            //初始化 HAL 库，完成中断分组

    //enable interrupt
    __set_PRIMASK(0);
    //System clock initialization
    clk_init(clock_src, clock_src_freq, clock_target_freq);  //配置系统时钟
    //disable interrupt
    __set_PRIMASK(1);

    rt_hw_systick_init();
```

```
        //Pin driver initialization is open by default
#ifdef RT_USING_PIN
        extern int rt_hw_pin_init(void);
        rt_hw_pin_init();                                    //初始化 PIN 设备
#endif

        //USART driver initialization is open by default
#ifdef RT_USING_SERIAL
        extern int rt_hw_usart_init(void);
        rt_hw_usart_init();                                  //初始化串口设备
#endif

}
```

RT-Thread 尽可能地编写了驱动层的代码，将系统级的配置统一放在 rtconfig.h 中进行，将板级独有的配置统一放在 board.h 中进行，因此只需要在 rtconfig.h 和 board.h 中配置相应的宏就能够配置相应的驱动，如 ADC、PWM、USART、SPI 等。在 RT-Thread Studio 中，rtconfig.h 中的配置项由 RT-Thread Settings 生成，board.h 中的配置只包含了一些示例，需要用户查阅相关资料。配置错误将导致驱动无法正常工作。

8）配置系统时钟（zonesion/common/BSP/drivers/drv_clk.c）

系统时钟的配置函数是 drv_clk.c 中 clk_init() 函数，该函数为各个硬件模块提供了工作时钟的基础。一般 clk_init() 函数是在 hw_board_init() 函数中调用的，hw_board_init() 函数位于 drv_common.c 中。RT-Thread Studio 生成的代码默认使用的是内部时钟，由于各家厂商的开发板使用的外部晶振频率不一定相同，因此需要根据具体的开发板来修改 clk_init() 函数，从而启动外部晶振。下面的代码给出了 clk_init() 函数的一种调用方式，这里使用的外部晶振频率为 8 MHz。

```
/******************************************************************************
 * 名称：clk_init(char *clk_source, int source_freq, int target_freq)
 * 功能：时钟源初始化
 * 参数：clk_source 表示时钟源，如 HSE 表示外部晶振；source_freq 表示外部晶振频率；target_freq
表示外部晶振的目标频率
 ******************************************************************************/
void clk_init(char *clk_source, int source_freq, int target_freq)
{
    if (strcmp(clk_source, "HSE") == 0)                      //外部晶振
    {
        RCC_OscInitTypeDef RCC_OscInitStruct = { 0 };
        RCC_ClkInitTypeDef RCC_ClkInitStruct = { 0 };

        //Configure the main internal regulator output voltage
        __HAL_RCC_PWR_CLK_ENABLE()
        ;
        __HAL_PWR_VOLTAGESCALING_CONFIG(PWR_REGULATOR_VOLTAGE_SCALE1);
        //Initializes the CPU, AHB and APB busses clocks
```

```
    RCC_OscInitStruct.OscillatorType = RCC_OSCILLATORTYPE_HSE;
    RCC_OscInitStruct.HSEState = RCC_HSE_ON;
    RCC_OscInitStruct.PLL.PLLState = RCC_PLL_ON;
    RCC_OscInitStruct.PLL.PLLSource = RCC_PLLSOURCE_HSE;
    RCC_OscInitStruct.PLL.PLLM = 8;
    RCC_OscInitStruct.PLL.PLLN = target_freq * 2;
    RCC_OscInitStruct.PLL.PLLP = RCC_PLLP_DIV2;
    RCC_OscInitStruct.PLL.PLLQ = 7;
    if (HAL_RCC_OscConfig(&RCC_OscInitStruct) != HAL_OK)
    {
        Error_Handler();
    }
    //Initializes the CPU, AHB and APB busses clocks
    RCC_ClkInitStruct.ClockType = RCC_CLOCKTYPE_HCLK | RCC_CLOCKTYPE_SYSCLK |
                            RCC_CLOCKTYPE_PCLK1 | RCC_CLOCKTYPE_PCLK2;
    RCC_ClkInitStruct.SYSCLKSource = RCC_SYSCLKSOURCE_PLLCLK;
    RCC_ClkInitStruct.AHBCLKDivider = RCC_SYSCLK_DIV1;
    RCC_ClkInitStruct.APB1CLKDivider = RCC_HCLK_DIV4;
    RCC_ClkInitStruct.APB2CLKDivider = RCC_HCLK_DIV2;

    if (HAL_RCC_ClockConfig(&RCC_ClkInitStruct, FLASH_LATENCY_5) != HAL_OK)
    {
        Error_Handler();
    }
    }
    else
    {
        system_clock_config(target_freq);
    }
}
```

9）实现操作系统节拍（zonesion/common/BSP/drivers/drv_common.c）

操作系统节拍也称为时钟节拍或滴答（Tick）时钟节拍。任何操作系统都需要提供一个时钟节拍，以供系统处理所有和时间有关的事件。时钟节拍的实现过程是：通过硬件定时器（Timer）实现周期性中断，在定时器中断时调用 rt_tick_increase()函数实现全局变量 rt_tick 的自加，从而实现时钟节拍。一般地，基于 Cortex-M 内核的微控制器直接使用内部的滴答定时器 Systick 实现时钟节拍。

时钟节拍由配置为中断触发模式的硬件定时器产生。当中断到来时，将调用一次 rt_tick_increase()函数，通知操作系统已经过去一个系统时钟。不同硬件定时器的中断实现往往不同，下面的代码以 STM32 系列微控制器为例介绍时钟节拍的配置。在初始化时钟节拍后，直接在 SysTick_Handler()函数（中断服务程序）中调用 rt_tick_increase()。rt_hw_systick_init()函数由 hw_board_init()函数调用，rt_hw_systick_init()函数使用 HAL 库完成系统时钟的配置，并设置系统时钟的中断优先级，这里的优先级是最高的。

```
/*SysTick configuration
void rt_hw_systick_init(void)
{
```

```
#if defined (SOC_SERIES_STM32H7)
    HAL_SYSTICK_Config((HAL_RCCEx_GetD1SysClockFreq()) / RT_TICK_PER_SECOND);
#else
    HAL_SYSTICK_Config(HAL_RCC_GetHCLKFreq() / RT_TICK_PER_SECOND);
#endif
    HAL_SYSTICK_CLKSourceConfig(SYSTICK_CLKSOURCE_HCLK);
    HAL_NVIC_SetPriority(SysTick_IRQn, 0, 0);
}
//中断服务程序
void SysTick_Handler(void)
{
    //进入中断
    rt_interrupt_enter();
    HAL_IncTick();
    rt_tick_increase();
    //离开中断
    rt_interrupt_leave();
}
```

1.4.2　开发设计与实践

本节在 ZI-ARMEmbed 上完成 RT-Thread 的移植。

1.4.3　开发步骤与验证

1.4.3.1　硬件部署

同 1.2.3.1 节。

1.4.3.2　系统移植

由于 02-Template 工程是由 RT-Thread Studio 自动生成的，具有一定的参考意义，因此本节在 02-Template 工程的基础上，针对实际的硬件进行 BSP 移植，从而构成一个完整的实例工程，便于在后文建立各个项目的代码。读者也可以直接导入已经创建好的 03-init 工程。

1）创建工程

（1）复制 02-Template 工程并命名为 03-init，将其复制到 RT-ThreadStudio/workspace 下。

（2）修改.cproject 第 215 行的目录路径，将 workspacePath 中 02-Template 修改为 03-init。.cproject 的代码请参考本书的配套资源（目录为"CH01-RT-Thread 开发基础/03-init/.cproject"）。.cproject 的部分代码如下：

```
<storageModule moduleId="refreshScope" versionNumber="2">
    <configuration configurationName="Debug">
        <resource resourceType="PROJECT" workspacePath="/03-init"/>
    </configuration>
</storageModule>
```

（3）修改.project 第 3 行和倒数第 3 行的工程名，将 name 中 02-Template 修改为 03-init。.project 的代码请参考本书的配套资源（目录为"CH01-RT-Thread 开发基础 /03-init/.project"）。.project 的部分代码如下：

```
<?xml version="1.0" encoding="UTF-8"?>
<projectDescription>
    <name>03-init</name>
    <comment />
    <projects>
</projects>
    <buildSpec>
        <buildCommand>
            <name>org.eclipse.cdt.managedbuilder.core.genmakebuilder</name>
            <triggers>clean,full,incremental,</triggers>
            <arguments>
            </arguments>
        </buildCommand>
        <buildCommand>
            <name>org.eclipse.cdt.managedbuilder.core.ScannerConfigBuilder</name>
            <triggers>full,incremental,</triggers>
            <arguments>
            </arguments>
        </buildCommand>
    </buildSpec>
    <natures>
        <nature>org.eclipse.cdt.core.cnature</nature>
        <nature>org.rt-thread.studio.rttnature</nature>
        <nature>org.eclipse.cdt.managedbuilder.core.managedBuildNature</nature>
        <nature>org.eclipse.cdt.managedbuilder.core.ScannerConfigNature</nature>
    </natures>
    <name>03-init</name>
    <linkedResources />
</projectDescription>
```

（4）修改.settings 目录下的 projcfg.ini 最后一行中的工程名，将 project_name 中的 02-Template 修改为 03-init。projcfg.ini 的代码请参考本书的配套资源（目录为"CH01-RT-Thread 开发基础/03-init/.settings /projcfg.ini"）。projcfg.ini 的部分代码如下：

```
#RT-Thread Studio Project Configuration
#Fri Sep 02 15:21:23 CST 2022
project_type=rtt
chip_name=STM32F407VE
cpu_name=None
target_freq=168
clock_source=hsi
dvendor_name=STMicroelectronics
rx_pin_name=PA10
rtt_path=repo/Extract/RT-Thread_Source_Code/RT-Thread/4.1.0
source_freq=0
csp_path=repo/Extract/Chip_Support_Packages/RealThread/STM32F4/0.2.2
```

```
sub_series_name=STM32F407
selected_rtt_version=4.1.0
cfg_version=v3.0
tool_chain=gcc
uart_name=uart1
tx_pin_name=PA9
rtt_nano_path=
output_project_path=C\:/RT-ThreadStudio/workspace
hardware_adapter=J-Link
project_name=03-init
```

（5）将.settings 目录中 4 个配置文件头修改为新的工程名，修改前后的工程名如图 1.42 所示。

（a）修改前的工程名

（b）修改后的工程名

图 1.42 修改前后的工程名

2）导入工程

（1）运行 RT-Thread Studio，选择菜单"文件"→"导入"，在弹出的"导入"对话框中选择"RT-Thread Studio 项目到工作空间中"，单击"下一步"按钮，在"导入项目"视图中将"选择根目录"设置为"RT-ThreadStudio\workspace\03-init"，"项目"下会显示所有的工程，勾选"03-init"，如图 1.43 和图 1.44 所示。

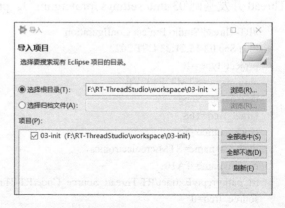

图 1.43 导入工程（一）　　　　图 1.44 导入工程（二）

（2）导入工程后的界面如图 1.45 所示。如果存在多个工程，则可通过双击工程名进行工程切换。

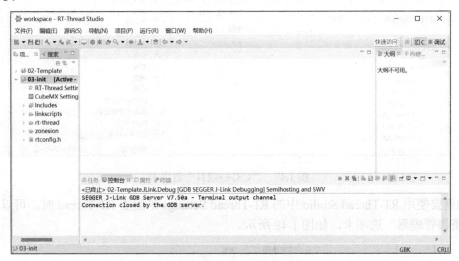

图 1.45　导入工程后的界面

3）调整工程结构

（1）在 03-init 工程根目录下，创建 zonesion 目录。

（2）在 zonesion 目录下，创建 app 目录、common 目录。在 app 目录下，创建 board 目录；在 common 目录下，创建 APL、BSP、DEV、DRV、LIB 目录。在 RT-Thread Studio 中，右键单击"03-init"，在弹出的右键菜单中选择"刷新"即可看到建立的目录，如图 1.46 所示。

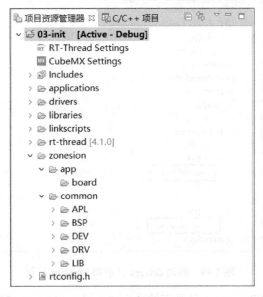

图 1.46　建立的目录

（3）工程视图。工程栏有两个选项卡："项目资源管理器"和"C/C++项目"。在"C/C++项目"选项卡中可以看到除编译文件之外的目录，如图 1.47 所示。

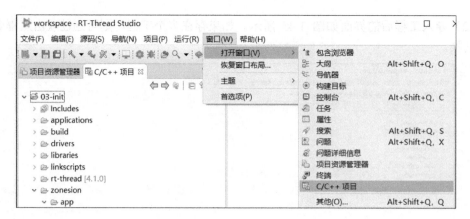

图 1.47 "C/C++项目"选项卡

当需要使用 RT-Thread Studio 中的 RT-Thread Settings 来配置 RT-Thread 时，可以切换到 "项目资源管理器"选项卡，如图 1.48 所示。

图 1.48 "项目资源管理器"选项卡

（4）将 03-init 工程根目录下的 drivers 目录移动到 common/BSP 目录下，将 libraries 目录移动到 common/LIB 目录下，如图 1.49 所示。

图 1.49 移动 drivers 目录和 libraries 目录

（5）将 common/BSP/drivers 目录下的 board.c、board.h、stm32f4xx_hal_conf.h 移动到 app/board 目录下。这三个文件的作用是配置 RT-Thread 和 STM32 系列微控制器的 HAL 库，因此不同的工程会有所不同，属于每个工程的独有文件。board.c、board.h、stm32f4xx_hal_conf.h 的位置如图 1.50 所示。

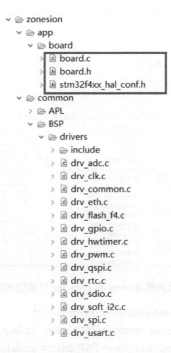

图 1.50　board.c、board.h、stm32f4xx_hal_conf.h 的位置

（6）将 applications/main.c 移动到 app/main.c，然后删除空的 applications 目录。

（7）在 zonesion 目录下创建 readme 目录，并在 readme 目录下新建 readme.txt 文件，该文件用于保存本工程的介绍和软件配置过程。

zonesion 目录的作用如下：

```
|--zonesion :
      |--app : 本工程独有的应用逻辑设计
            |--board : 板级驱动与配置
      |--common : 本工程硬件电路板所有项目可共用的应用软件包与驱动文件
            |--APL : 应用软件包（如 Modbus 协议栈、通信模块应用协议等）
            |--BSP : 板级支持包，包含 MCU 官方定义的和自定义的各项外设驱动
            |--DEV : 一些设备支持文件（如 MCU 外设的 MSP 层实现）
            |--DRV : 硬件设备驱动（比如传感器、控制器）
            |--LIB : STM32 官方的驱动库
      |--readme
```

4）修改头文件路径

由于前文对工程目录进行了调整，因此需要重新配置工程，将刚刚创建的目录都加入头文件搜索路径（如 zonesion/app、zonesion/app/board、zonesion/common/APL、zonesion/common/BSP、zonesion/common/DEV、zonesion/common/DRV、zonesion/common/LIB），下面介绍具体的操作步骤。

（1）右键单击工程名"03-init"，在弹出的右键菜单中选择"属性"，可弹出"03-init 的属性"对话框，在该对话框中选择"C/C++构建"→"设置"，可以看到使用的 C 的交叉编译器 GNU ARM Cross C Compiler，在 includes 中修改原来的 drivers 和 libraries 目录的相对路径（见图 1.51），代码如下：

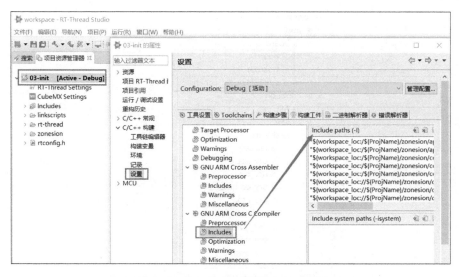

图 1.51　修改原来的 drivers 和 libraries 目录的相对路径

"${workspace_loc://${ProjName}/zonesion/common/BSP/drivers}"
"${workspace_loc://${ProjName}/zonesion/common/BSP/drivers//include}"
"${workspace_loc://${ProjName}/zonesion/common/BSP/drivers//include//config}"
"${workspace_loc://${ProjName}/zonesion/common/LIB/libraries//CMSIS//Device//ST//STM32F4xx//Include}"
"${workspace_loc://${ProjName}/zonesion/common/LIB/libraries//CMSIS//Include}"
"${workspace_loc://${ProjName}/zonesion/common/LIB/libraries//CMSIS//RTOS//Template}"
"${workspace_loc://${ProjName}/zonesion/common/LIB/libraries//STM32F4xx_HAL_Driver//Inc}"
"${workspace_loc://${ProjName}/zonesion/common/LIB/libraries//STM32F4xx_HAL_Driver//Inc//Legacy}"

（2）增加应用程序和板级配置文件夹的搜索路径 zonesion/app、zonesion/app/board。增加搜索路径如图 1.52 和图 1.53 所示。

图 1.52　增加搜索路径（一）

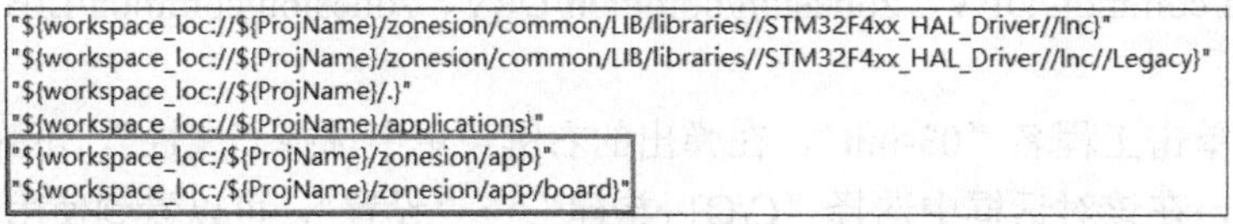

"${workspace_loc://${ProjName}/zonesion/common/LIB/libraries//STM32F4xx_HAL_Driver//Inc}"
"${workspace_loc://${ProjName}/zonesion/common/LIB/libraries//STM32F4xx_HAL_Driver//Inc//Legacy}"
"${workspace_loc://${ProjName}/.}"
"${workspace_loc://${ProjName}/applications}"
"${workspace_loc:/${ProjName}/zonesion/app}"
"${workspace_loc:/${ProjName}/zonesion/app/board}"

图 1.53　增加搜索路径（二）

（3）增加驱动相关的头文件路径，即 zonesion/common/APL、zonesion/common/DEV、

zonesion/common/DRV 的搜索路径,如图 1.54 所示。

```
"${workspace_loc:/${ProjName}/zonesion/common/APL}"
"${workspace_loc:/${ProjName}/zonesion/common/DEV}"
"${workspace_loc:/${ProjName}/zonesion/common/DRV}"
"${workspace_loc://${ProjName}/zonesion/common/BSP/drivers}"
"${workspace_loc://${ProjName}/zonesion/common/BSP/drivers//include}"
"${workspace_loc://${ProjName}/zonesion/common/BSP/drivers//include//config}"
```

图 1.54　增加驱动相关的头文件路径

(4)删除多余的头文件搜索路径,即删除 " "${workspace_loc://${ProjName}/applications}""。

(5)再次编译,可发现编译通过,说明头文件路径配置正确。

5)修改系统时钟

(1)由于 RT-Thread 生成的工程在默认情况下使用的是内部晶振,因此需要修改代码来使用外部晶振,并替换本工程中的 "03-init/zonesion/common/BSP/drivers/drv_clk.c"。

(2)由于 clk_init()函数是在 hw_board_init()函数中引用的,hw_board_init()函数是在 rt_hw_board_init()函数引用的,而 hw_board_init()函数中的入口参数为 BSP_CLOCK_SOURCE、BSP_CLOCK_SOURCE_FREQ_MHZ、BSP_CLOCK_SYSTEM_FREQ_MHZ(见图 1.55),这三个宏的定义在 "zonesion/app/board.h" 中。

```
   drv_clk.c    board.c    drv_common.c

10
11  #include <rtthread.h>
12  #include <board.h>
13  #include <drv_common.h>
14
15  RT_WEAK void rt_hw_board_init()
16  {
17      extern void hw_board_init(char *clock_src, int32_t clock_src_freq, int32_t clock_target_freq)
18
19      /* Heap initialization */
20  #if defined(RT_USING_HEAP)
21      rt_system_heap_init((void *) HEAP_BEGIN, (void *) HEAP_END);
22  #endif
23
24      hw_board_init(BSP_CLOCK_SOURCE, BSP_CLOCK_SOURCE_FREQ_MHZ, BSP_CLOCK_SYSTEM_FREQ_MHZ);
25
```

图 1.55　hw_board_init()函数中的入口参数

按照实际的外部晶振修改参数即可,修改前的 BSP_CLOCK_SOURCE 如图 1.56 所示。

```
   drv_clk.c    board.c    drv_common.c    board.h

43  /*----------------------- CLOCK CONFIG BEGIN -----------------------*/
44
45  #define BSP_CLOCK_SOURCE            ("HSI")
46  #define BSP_CLOCK_SOURCE_FREQ_MHZ   ((int32_t)0)
47  #define BSP_CLOCK_SYSTEM_FREQ_MHZ   ((int32_t)168)
48
```

图 1.56　修改前的 BSP_CLOCK_SOURCE

修改前的 BSP_CLOCK_SOURCE 如图 1.57 所示。

```
   drv_clk.c    board.c    drv_common.c    board.h    board.h

43  /*----------------------- CLOCK CONFIG BEGIN -----------------------*/
44
45  #define BSP_CLOCK_SOURCE            ("HSE")
46  #define BSP_CLOCK_SOURCE_FREQ_MHZ   ((int32_t)8)
47  #define BSP_CLOCK_SYSTEM_FREQ_MHZ   ((int32_t)168)
```

图 1.57　修改后的 BSP_CLOCK_SOURCE

6)修改调试配置

单击工具栏中的 " ⚙ "(调试配置)按钮,可打开"配置工程"对话框,选择 "SVD Path"

选项卡，将"File path"中的 0.2.0 改为 0.2.2，如图 1.58 所示。

图 1.58　将"File path"中的 0.2.0 改为 0.2.2

1.4.3.3　程序调试

1）导入工程

（1）前文已经完成了工程的创建，读者可直接将 03-init 工程复制到 RT-Thread Studio。

（2）如果 RT-Thread Studio 中存在多个工程，可通过双击工程名来切换工程。

2）编译工程

（1）单击工具栏中的" 🔧 "（构建）按钮完成工程的编译。

（2）如果控制台窗口提示编译完成且没有报错，则表示工程编译成功。

3）下载工程

（1）单击工具栏中的" 📥 "（下载）按钮可将编译后的工程下载到 ZI-ARMEmbed。

（2）如果控制台窗口提示下载完成且没有报错，则表示工程下载成功。

（3）工程下载完成后，按下 ZI-ARMEmbed 上的复位键即可运行新的工程，并在 MobaXterm 串口终端 FinSH 控制台中看到工程启动时的信息。

4）调试工程

（1）单击工具栏中的" 🐞 "（调试）按钮可启动调试功能。

（2）在程序合适的位置设置断点，单步对程序进行调试，观察变量的变化和寄存器的变化，可加深对程序的理解。

（3）调试完后，单击工具栏中的" ⬛ "（停止）按钮可退出调试模式。

1.4.3.4　验证效果

（1）关闭 RT-Thread Studio，拔掉仿真器，按下 ZI-ARMEmbed 上的电源按键重新上电。

（2）本工程根据实际的硬件对工程结构进行了调整，完成了 RT-Thread 的移植，MobaXterm 串口终端 FinSH 控制台会默认每秒显示一次"Hello RT-Thread!"消息。

1.4.4　小结

本节介绍了 RT-Thread 的移植，涉及的内容包括系统的启动流程、上下文的切换、线程的初始化等。通过本节的学习，读者可完成 RT-Thread 的移植。

1.5 FinSH 控制台的应用

开发人员与嵌入式系统的交互工具和命令在嵌入式系统的开发中是至关重要的。这些工具和命令允许开发人员配置、调试、测试和监控嵌入式系统。以下是一些常用的工具和命令：

（1）串口终端工具：通过串口连接嵌入式系统后，开发人员可以利用串口终端工具（如 PuTTY、Tera Term、Minicom 等）执行命令、查看输出，并与嵌入式系统进行交互。通常，串口终端工具需要配置串口的波特率、数据位、校验位和停止位等。

（2）SSH（Secure Shell）：如果嵌入式系统支持 SSH，开发人员就可以通过 SSH 连接到嵌入式系统，进行安全的远程终端会话。SSH 通常需要使用用户名和密码或通过 SSH 密钥进行身份验证。

（3）Telnet：Telnet 是一种网络协议，允许开发人员通过网络连接到嵌入式系统。Telnet 类似于 SSH，但不提供加密功能。

（4）命令行界面（CLI）：嵌入式系统通常会提供一个命令行界面，允许开发人员执行系统命令、配置参数和查看状态信息。通过命令行界面可进行文件操作、网络配置、进程管理等。

（5）GDB（GNU Debugger）：GDB 是一种功能强大的调试器，可用于调试嵌入式应用程序。通过 GDB，开发人员可以设置断点、单步执行代码、检查变量值和调试嵌入式系统。

（6）JTAG 和 SWD：是用于硬件调试的接口。开发人员可以使用 JTAG 或 SWD 连接到嵌入式处理器的调试接口，从而进行硬件调试、故障排除。

（7）FinSH 控制台（Friendly Interactive Shell）控制台：是一个轻量级的嵌入式系统命令行 Shell，通常用于嵌入式系统开发。FinSH 控制台允许开发人员与嵌入式系统进行交互，如运行命令、查看系统状态、进行调试等操作，以便轻松地开发和调试嵌入式系统。

本节的要求如下：

⊃ 了解自定义 FinSH 控制台命令的基本原理。

⊃ 掌握由 RT-Thread Studio 的基础工程拓展其他工程的方法。

⊃ 掌握 RT-Thread 和自定义 FinSH 控制台命令的应用。

1.5.1　原理分析

1.5.1.1　FinSH 控制台简介

FinSH 控制台是 RT-Thread 的命令行组件，提供了一套供用户在命令行中调用的操作接口，主要用于调试或查看系统信息。FinSH 控制台可以使用串口、以太网、USB 等与 PC 进行通信。FinSH 控制台与 PC 的连接如图 1.59 所示。

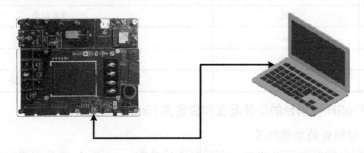

图 1.59　FinSH 控制台与 PC 的连接

用户在控制终端（如 PC）输入命令，控制终端通过串口、USB、网络等方式将命令传给设备里的 FinSH 控制台，FinSH 控制台会读取设备输入命令，解析并自动扫描内部函数表，寻找对应函数名，执行函数后输出响应，响应通过原路返回，将结果显示在控制终端上。

FinSH 控制台支持自动补全、查看历史命令等功能，通过键盘上的按键可以很方便地使用这些功能，FinSH 控制台支持的按键功能如表 1.6 所示。

表 1.6　FinSH 控制台支持的按键功能

按　　键	功 能 描 述
Tab	当没有输入任何字符时按下 Tab 键将会显示当前系统支持的所有命令。若已经输入部分字符时按下 Tab 键，则会查找匹配的命令，也会按照文件系统的当前目录下的文件名进行补全，并可以继续输入，多次补全
↑、↓	上下翻阅最近输入的历史命令
Backspace	删除符
←、→	向左或向右移动光标

1.5.1.2　FinSH 控制台的内置命令

RT-Thread 默认地内置了一些 FinSH 控制台命令，在 FinSH 控制台中输入 help 后按下 Enter 键或者 Tab 键，就可以显示当前系统支持的所有命令。常用的 FinSH 控制台内置命令如表 1.7 所示。

表 1.7　常用的 FinSH 控制台内置命令

序　　号	命　　令	描　　述
1	version	显示 RT-Thread 版本信息
2	list_thread	列出所有的线程状态
3	list_sem	列出系统中所有的信号量状态
4	list_event	列出系统中所有的事件状态
5	list_mutex	列出系统中所有的互斥量状态
6	list_mailbox	列出系统中所有的邮箱状态
7	list_msgqueue	列出系统中所有的消息队列状态
8	list_timer	列出系统中所有的定时器状态
9	list_device	列出系统中所有的设备状态
10	exit	退出 FinSH 控制台
11	help	显示帮助信息
12	ps	显示线程信息
13	time	显示时间信息
14	free	显示系统内存的使用状态

1.5.1.3　FinSH 控制台的功能配置和自定义 FinSH 控制台命令

1）FinSH 控制台的功能配置

FinSH 控制台的功能可以裁减，宏配置选项是在 rtconfig.h 文件中进行的，具体配置项如表 1.8 所示。在实际的开发中，用户可以根据需求进行配置。

表 1.8　FinSH 控制台的功能配置项

宏 定 义	取 值 类 型	描 述	默 认 值
#define RT_USING_FINSH	无	使能 FinSH 控制台	开启
#define FINSH_THREAD_NAME	字符串	FinSH 控制台的线程名字	tshell
#define FINSH_USING_HISTORY	无	打开历史回溯功能	开启
#define FINSH_HISTORY_LINES	整型数据	可回溯的历史命令行数	5
#define FINSH_USING_SYMTAB	无	可在 FinSH 控制台使用的符号表	开启
#define FINSH_USING_DESCRIPTION	无	给每个 FinSH 控制台的符号添加一段描述	开启
#define FINSH_USING_MSH	无	使能 MSH 模式	开启
#define FINSH_USING_MSH_ONLY	无	只使用 MSH 模式	开启
#define FINSH_ARG_MAX	整型数据	输入参数的最大数量	10
#define FINSH_USING_AUTH	无	使能权限验证	关闭
#define FINSH_DEFAULT_PASSWORD	字符串	权限密码验证	关闭

2）自定义 FinSH 控制台命令

除了 FinSH 控制台内置的命令，FinSH 控制台还提供了多个宏接口来导出自定义命令，导出的命令可以直接在 FinSH 控制台中执行。RT-Thread 同时支持自定义 MSH 命令和自定义 C-Style 命令，自定义 MSH 命令的使用更加广泛，如果没有特殊说明，本书中所说的自定义 FinSH 控制台命令都是 MSH 命令。

自定义的 FinSH 控制台命令可以在 MSH 模式下运行，将一个命令导出到 MSH 模式可以使用如下宏接口如下：

```
/****************************************************************************
 * 名称：MSH_CMD_EXPORT
 * 功能：自定义 MSH 命令导出
 * 参数：name 表示要导出的命令；desc 表示导出命令的描述
 ****************************************************************************/
MSH_CMD_EXPORT(name, desc);
```

宏接口 MSH_CMD_EXPORT()既可以导出有参数的命令，也可以导出无参数的命令。在导出无参数命令时，宏接口 MSH_CMD_EXPORT()的入参为 void。示例如下：

```
#include <rtthread.h>
void hello(void)
{
    rt_kprintf("hello RT-Thread!\n");
}
MSH_CMD_EXPORT(hello , say hello to RT-Thread);
```

在导出有参数的命令时，宏接口 MSH_CMD_EXPORT()的入参为 int argc 和 char**argv。argc 表示参数的个数，argv 表示指向命令行参数字符串指针的数组指针。示例如下：

```
#include <rtthread.h>
static void atcmd(int argc, char**argv)
{
    if (argc < 2)
```

```
    {
        rt_kprintf("Please input'atcmd <server|client>'\n");
        return;
    }
    if (!rt_strcmp(argv[1], "server"))
    {
        rt_kprintf("AT server!\n");
    }
    else if (!rt_strcmp(argv[1], "client"))
    {
        rt_kprintf("AT client!\n");
    }
    else
    {
        rt_kprintf("Please input'atcmd <server|client>'\n");
    }
}
MSH_CMD_EXPORT(atcmd, atcmd sample: atcmd <server|client>);
```

1.5.2　开发设计与实践

1.5.2.1　硬件设计

要实现自定义 FinSH 控制台命令，首先要了解如何定义 FinSH 控制台命令并将其添加到命令列表中；其次要让命令效果更加明显。本节使用 FinSH 控制台来控制 LED，将对 LED 的控制转化为对 GPIO 的控制，即控制 GPIO 输出高电平和低电平，如图 1.60 所示，从而将自定义的 FinSH 控制台命令和 LED 的控制结合起来实现联动控制。

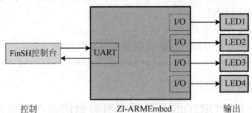

图 1.60　将对 LED 的控制转化为对 GPIO 的控制

ZI-ARMEmbed 和 LED 连接如图 1.61 所示，图中 LED 的一端连接 3.3 V 的电源，另一端通过电阻连接 ZI-ARMEmbed 中的 STM32F407。当 PE0、PE1、PE2 和 PE3 为高电平时，LED 两端电压相同，无法形成压降，LED 熄灭。反之当 PE0、PE1、PE2 和 PE3 为低电平时，LED 的两端形成压降，LED 点亮。

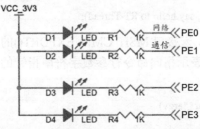

图 1.61　ZI-ARMEmbed 和 LED 的连接

1.5.2.2 软件设计

软件设计流程如图 1.62 所示。

（1）在 main 线程中进行 LED 引脚的初始化，配置相关 GPIO 引脚的模式和初始电平。

（2）编写自定义 FinSH 控制台命令函数，当输入的 FinSH 控制台命令是"ledCtrl on"时点亮 4 个 LED，当输入的 FinSH 控制台命令是"ledCtrl off"时熄灭 4 个 LED。

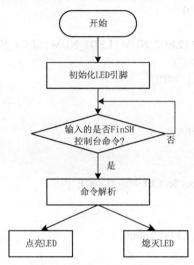

图 1.62　软件设计流程

1.5.2.3 功能设计与核心代码设计

通过原理学习可知，要实现自定义 FinSH 控制台命令，就需要使用 RT-Thread 提供的接口将自定义的 FinSH 控制台命令添加到命令列表中，这样在启动 FinSH 控制台之后就可以使用自定义的 FinSH 控制台命令了。

1）主函数（zonesion/app/main.c）

主函数的主要工作是调用 led_pin_init()函数，完成引导工作。代码如下：

```
/***********************************************************************
 * 名称：main()
 * 功能：调用 led_pin_init()函数，完成引导工作
 ***********************************************************************/
#include "drv_led.h"
int main(void)
{
    led_pin_init();                                              //LED 引脚初始化
    return 0;
}
```

2）自定义 FinSH 控制台命令函数（zonesion/app/finsh_ex.c）

自定义 FinSH 控制台命令函数实现 LED 的开关控制，代码如下：

```
/***********************************************************************
 * 名称：ledCtrl()
 * 功能：自定义 FinSH 控制台命令
```

```
    * 参数：argc 表示参数个数；argv 表示指向参数字符串指针的数组指针
    **********************************************************************/
#include <rtthread.h>
#include "drv_led.h"
int ledCtrl(int argc, char **argv)
{
    if(!rt_strcmp(argv[1], "on"))
    {
        led_ctrl(LED1_NUM | LED2_NUM | LED3_NUM | LED4_NUM);
        rt_kprintf("LED ON!\n");
    } else if(!rt_strcmp(argv[1], "off"))
    {
        led_ctrl(0);
        rt_kprintf("LED OFF!\n");
    }
    else
        rt_kprintf("Please input 'ledCtrl <on|off>'\n");
    return 0;
}
MSH_CMD_EXPORT(ledCtrl, LED Control);
```

1.5.3 开发步骤与验证

1.5.3.1 硬件部署

同 1.2.3.1 节。

1.5.3.2 创建工程

（1）本节在 03-init 工程的基础上增加了控制 LED 的 FinSH 控制台命令代码。增加的代码如下：

```
zonesion/app/main.c
zonesion/app/main.h
zonesion/app/finsh_cmd/ finsh_ex.c
zonesion/common/DRV/drv_led/drv_led.c
zonesion/common/DRV/drv_led/drv_led.h
```

（2）导入 04-FinSH 控制台工程。

（3）双击 "04-FinSH" → "RT-Thread Settings"，可弹出 "RT-Thread Settings" 窗口，如图 1.63 所示。本节使用 GPIO 来控制 LED，所以需要打开 PIN 设备，在工程中该设备已经默认打开（在实际中彩色图标表示已打开），所以这里不需要进行任何设置。

（4）添加 LED 驱动头文件路径。右键单击工程名 "04-FinSH"，在弹出的右键菜单中选择 "属性" 可弹出 "04-FinSH 的属性" 对话框，在该对话框中选择 "C/C++构建" → "设置"，可以看到使用的 C 的交叉编译器为 GNU ARM Cross C Compiler，在 includes 中添加 LED 驱动头文件的路径 ""${workspace_loc:/${ProjName}/zonesion/common/DRV/drv_led}""，如图 1.64 所示。

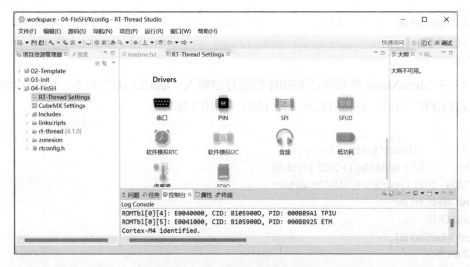

图 1.63　"RT-Thread Settings"窗口

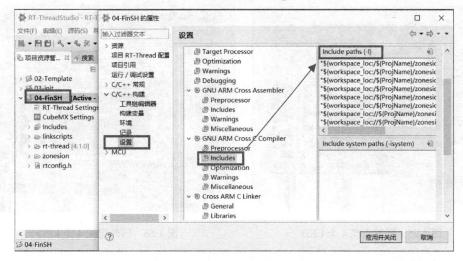

图 1.64　添加 LED 驱动头文件路径

至此就完成了工程的修改。

1.5.3.3　项目调试与验证效果

（1）运行 RT-Thread Studio，选择菜单"文件"→"导入"，在弹出的"导入"对话框中选择"RT-Thread Studio 项目到工作空间中"，单击"下一步"按钮，在"导入项目"视图中将"选择根目录"设置为"RT-ThreadStudio\workspace\05-FinSH"，在"项目"下会显示所有的工程，勾选"04-FinSH"后单击"完成"按钮。

（2）关闭 RT-Thread Studio，拔掉仿真器，按下 ZI-ARMEmbed 上的电源按键重新上电。

（3）在 MobaXterm 串口终端 FinSH 控制台中输入"ledCtrl on"命令，FinSH 控制台输出"LED ON!"信息，并同时点亮 4 个 LED（见图 1.65）。

```
       \ | /
     - RT -     Thread Operating System
       / | \     4.1.0 build Oct 13 2022 17:38:30
     2006 - 2022 Copyright by RT-Thread team
```

```
msh >ledCtrl on
LED ON!
msh >
```

（4）在 MobaXterm 串口终端 FinSH 控制台中输入"ledCtrl off"命令，FinSH 控制台输出"LED OFF"信息，并同时熄灭 4 个 LED（见图 1.66）。

```
  \ | /
- RT -      Thread Operating System
 / | \      4.1.0 build Oct 13 2022 17:38:30
 2006 - 2022 Copyright by RT-Thread team
msh >ledCtrl on
LED ON!
msh >ledCtrl off
LED OFF!
msh >
```

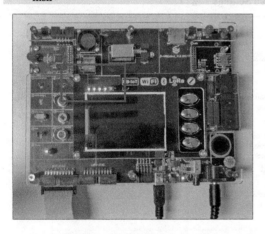

图 1.65　同时点亮 4 个 LED

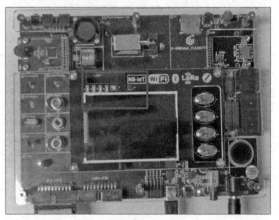

图 1.66　同时熄灭 4 个 LED

1.5.4　小结

本节介绍了 FinSH 控制台的应用，以及从 RT-Thread Studio 基础工程拓展其他工程的方法。通过本节的学习，读者可结合自定义 FinSH 控制台命令和 RT-Thread 进行简单的应用开发。

第 2 章
RT-Thread 内核开发技术

2.1 RT-Thread 线程管理应用开发

线程管理是多任务操作系统的核心功能之一。线程管理涉及线程的创建、调度、同步和销毁等操作。线程是一个进程内的独立执行流,多线程的使用可以提高系统的并发性和实时性。有效的线程管理可以提高系统的性能和并发处理能力。

本节的要求如下:

- 了解线程的基本概念。
- 学习线程调度的机制。
- 掌握线程管理方式。
- 掌握 RT-Thread 线程管理的应用。

2.1.1 原理分析

2.1.1.1 线程管理简介

在日常生活中,要解决一个大问题,一般会将它分解成多个简单、容易解决的小问题,小问题逐个被解决后,大问题也就随之被解决了。在多线程操作系统中,同样也需要把一个复杂的应用分解成多个小的、可调度的、序列化的程序单元(模块)。合理划分任务并正确执行各个子任务的方式,能够提高系统的性能并满足实时性的要求。例如,嵌入式系统通常需要通过传感器采集数据、通过显示屏将数据显示出来。在多线程操作系统中,可以将嵌入式系统的任务分为两个子任务,即接收传感器数据和输出显示,如图 2.1 所示,一个子任务不间断地读取传感器数据,并将数据写到共享内存中,另一个子任务周期性地从共享内存中读取数据并将其显示到显示屏上。

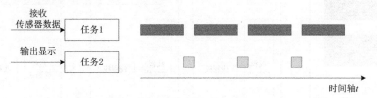

图 2.1 嵌入式系统的两个子任务

在 RT-Thread 中,与子任务对应的程序实体就是线程。线程是 RT-Thread 中最基本的调度单位,它不仅描述了一个任务执行的运行环境,还描述了这个任务的优先级。重要的任务可设置相对较高的优先级，不重要的任务可以设置较低的优先级，不同的任务还可以设置相同的优先级，轮流运行。

2.1.1.2 线程管理有关概念

（1）线程的创建和销毁。通过操作系统提供 API 可创建线程，如 pthread_create（POSIX 系统）或 CreateThread（Windows 系统）。销毁线程通常是由线程本身完成的，也可以在其他线程中调用特定 API 来终止目标线程。

（2）线程调度。操作系统通过调度算法来决定哪个线程在何时执行，这通常涉及线程的优先级、时间片分配和上下文切换。在切换线程时，需要保存和还原线程的上下文，包括寄存器状态和栈。

（3）线程的同步和互斥。互斥量用于保护临界区，防止多个线程同时访问共享资源，防止竞争条件（Race Conditions）。信号量用于控制多个线程的并发访问，限制资源的可用性。条件变量用于线程之间的协作，一个线程等待另一个线程发出通知。

（4）线程通信。①消息队列：线程可以通过消息队列相互通信，发送和接收消息。②共享内存：多个线程可以共享同一块内存区域，以进行数据交换。③管道和套接字：线程可以使用管道和套接字与其他进程进行通信。

（5）线程优先级。每个线程都具有一个优先级，用于决定该线程在调度时的执行顺序。高优先级的线程通常会被更早地调度执行。

（6）线程状态。线程可以处于不同的状态，如运行、就绪、阻塞等，具体的状态由操作系统管理。

（7）线程安全性。线程安全性是指多线程环境下，代码能够正确处理并发操作，不会导致数据损坏或不一致，编写线程安全的代码是至关重要的。

（8）线程池。线程池可以维护一组预创建的线程，以处理任务队列中的工作。线程池可以提高线程的重用性，减少线程创建和销毁的开销。

2.1.1.3 线程调度机制

RT-Thread 线程管理的主要功能是对线程进行管理和调度。RT-Thread 中的线程可分为两类：系统线程和用户线程。系统线程是由 RT-Thread 内核创建的，用户线程是由应用程序创建的。这两类线程在创建时都会从内核对象容器中分配线程对象，当销毁时都会删除对象容器中的线程对象。每个线程都有一些基本的属性，如线程控制块、线程栈、入口函数等。RT-Thread 的线程管理如图 2.2 所示。

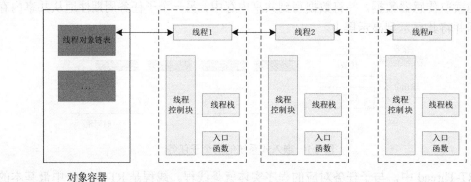

图 2.2　RT-Thread 线程管理

RT-Thread 的线程调度器是抢占式的，主要的工作就是从就绪的线程列表中查找优先级最高的线程，保证优先级最高的线程能够被运行，优先级最高的线程一旦就绪，总能得到 CPU 的使用权。在某个线程的运行过程中，当某个优先级更高的线程满足运行条件时，正在运行的线程的 CPU 使用权就会被剥夺，优先级高的线程可立刻得到 CPU 使用权。如果中断服务程序使一个优先级高的线程满足运行条件，则在中断服务程序完成时，被中断的线程将被挂起，先运行优先级高的线程。当线程调度器调度线程时，先将当前线程的上下文保存起来；当切回到原来的线程时，线程调度器将恢复该线程的上下文。

2.1.1.4　线程管理方式

图 2.3 所示为线程的相关操作，包含创建/初始化线程、启动线程、运行线程、删除/脱离线程。通过 rt_thread_create()函数可创建一个动态线程，通过 rt_thread_init()函数可初始化一个静态线程。动态线程与静态线程的区别是：动态线程是系统自动从动态内存上分配栈空间与线程句柄的（初始化内存后才能使用 rt_thread_create()函数创建动态线程），静态线程是由用户分配栈空间与线程句柄的。

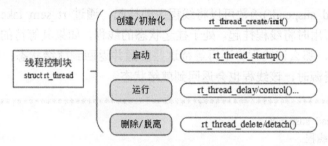

图 2.3　RT-Thread 线程管理方式

1）创建线程

一个线程要成为可执行的对象，就必须由操作系统的内核来为它创建一个线程。通过下面的函数可创建一个动态线程：

```
/******************************************************************************
 * 名称：rt_thread_create()
 * 功能：创建线程
 * 参数：name 表示线程的名称；entry 表示线程入口函数；parameter 表示线程入口函数参数；stack_size
表示线程栈大小，单位是字节；priority 表示线程的优先级；tick 表示线程的时间片大小
 * 返回：thread 表示线程创建成功时返回的线程句柄；RT_NULL 表示线程创建失败
 ******************************************************************************/
rt_thread_t rt_thread_create(const char* name, void (*entry)(void* parameter), void* parameter,
                    rt_uint32_t stack_size, rt_uint8_t priority, rt_uint32_t tick);
```

2）删除线程

对于由 rt_thread_create()函数创建的线程，在不需要使用该线程或者该线程运行出错时，可以使用下面的函数来删除线程：

```
/******************************************************************************
 * 名称：rt_thread_delete()
 * 功能：删除线程
 * 参数：thread 表示要删除的线程句柄
```

```
 * 返回：RT_EOK 表示删除线程成功；-RT_ERROR 表示删除线程失败
 *****************************************************************************/
rt_err_t rt_thread_delete(rt_thread_t thread);
```

3）启动线程

创建（初始化）的线程状态处于初始状态，并未进入就绪线程的调度队列。在线程初始化或创建成功时调用下面的函数可以令线程进入就绪态：

```
/*****************************************************************************
 * 名称：rt_thread_startup()
 * 功能：启动线程
 * 参数：thread 表示线程句柄
 * 返回：RT_EOK 表示线程启动成功；-RT_ERROR 表示线程启动失败
 *****************************************************************************/
rt_err_t rt_thread_startup(rt_thread_t thread);
```

4）挂起线程

通过 rt_thread_suspend()函数可以将线程主动挂起，通过 rt_sem_take()、rt_mb_recv()函数可以在资源不可用时将线程挂起。处于挂起状态的线程，如果其等待的资源超时（超过其设定的等待时间），那么该线程将不再等待这些资源并返回到就绪状态；当其他线程释放掉该线程所等待的资源时，该线程也会返回到就绪状态。

```
/*****************************************************************************
 * 名称：rt_thread_suspend()
 * 功能：挂起线程
 * 参数：thread 表示线程句柄
 * 返回：RT_EOK 表示线程挂起成功；-RT_ERROR 表示线程挂起失败
 *****************************************************************************/
rt_err_t rt_thread_suspend (rt_thread_t thread);
```

5）恢复线程

恢复线程就是让挂起的线程重新进入就绪状态，并将其放入就绪线程的调度队列。如果被恢复线程在所有的就绪态线程中位于最高优先级链表的第一位，那么系统将进行线程上下文的切换。通过下面的函数可恢复线程：

```
/*****************************************************************************
 * 名称：rt_thread_resume()
 * 功能：恢复线程
 * 参数：thread 表示线程句柄
 * 返回：RT_EOK 表示线程恢复成功；-RT_ERROR 表示线程恢复失败
 *****************************************************************************/
rt_err_t rt_thread_resume (rt_thread_t thread);
```

2.1.2　开发设计与实践

2.1.2.1　硬件设计

本节的硬件设计同 1.5.2.1 节。

2.1.2.2 软件设计

软件设计流程如图 2.4 所示。

（1）创建 LED 线程，并且在 main 函数中完成线程的初始化工作。

（2）编写线程入口函数，在线程入口函数中完成 LED 引脚的初始化工作。

（3）当在 MobaXterm 串口终端 FinSH 控制台中输入"threadManage start"命令时，可启动 LED 线程，并进入线程入口函数中循环点亮 4 个 LED 灯；当输入"threadManage suspend"命令时，可挂起线程，4 个 LED 停止循环点亮；当输入"threadManage resume"命令时，可恢复线程，4 个 LED 继续循环点亮；当输入"threadManage delete"命令时，可删除线程，4 个 LED 停止循环点亮，并且在系统中删除 LED 线程。

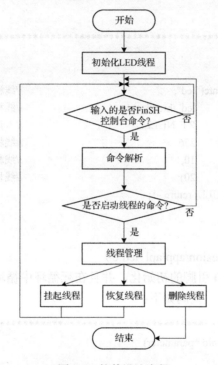

图 2.4 软件设计流程

2.1.2.3 功能设计与核心代码设计

通过原理学习可知，要实现 LED 线程的管理，就需要使用 RT-Thread 提供的线程管理接口进行线程的初始化，并编写相应的线程入口函数。线程管理使用的是 FinSH 控制台命令，需要实现对应的控制命令，并将实现的命令添加到 FinSH 控制台命令列表中，这样在启动 FinSH 控制台后就可以使用自定义的 FinSH 控制台命令了。

1）主函数（zonesion/app/main.c）

主函数的主要工作是调用 led_thread_init()函数，完成引导工作。代码如下：

```
/*******************************************************************************
 * 名称：main()
 * 功能：调用 led_thread_init()函数，完成引导工作
 *******************************************************************************/
#include "apl_led.h"
```

```
int main(void)
{
    led_thread_init();                                              //LED 线程初始化
    return 0;
}
```

2）线程初始化函数（zonesion/app/apl_led.c）

线程初始化函数实现线程的创建，包括定义线程名称、线程入口函数，设置线程内存大
小、线程优先级、线程时间片等参数。代码如下：

```
/**********************************************************************************
* 名称：led_thread_init()
* 功能：线程初始化
**********************************************************************************/
int led_thread_init(void)
{
    led_thread = rt_thread_create("led",                           //线程名称
                                  led_thread_entry,                //线程入口函数
                                  RT_NULL,                         //入口函数参数
                                  256,                             //线程内存大小
                                  10,                              //线程优先级
                                  20);                             //线程时间片
    if(led_thread == RT_NULL) return -1;
    return 0;
}
```

3）线程入口函数（zonesion/app/apl_led.c）

线程入口函数实现 LED 引脚的初始化，并且在死循环中循环点亮 LED。在线程启动后
即可进入线程入口函数运行。代码如下：

```
/**********************************************************************************
* 名称：led_thread_entry(void *parameter)
* 功能：线程入口函数
* 参数：parameter 表示入口参数，在创建线程时传入
**********************************************************************************/
static void led_thread_entry(void *parameter)
{
    (void)parameter;
    unsigned char ledState = 0x01;
    led_pin_init();                                                //LED 引脚初始化
    while(1)
    {
        if(suspendFlag == 1)
        {
            rt_thread_suspend(led_thread);                         //挂起线程
            rt_schedule();
            suspendFlag = 0;
        }
        led_ctrl(ledState);
```

```
                if(ledState < 0x08)
                    ledState <<= 1;
                else
                    ledState = 0x01;
                rt_thread_mdelay(500);                            //延时 500 ms
        }
}
```

4）自定义 FinSH 控制台命令函数（zonesion/app/finsh_cmd/finsh_ex.c）

自定义 FinSH 控制台命令函数实现 LED 线程管理的命令，包括线程的创建、挂起、恢复及删除，可以在 FinSH 控制台中使用命令对线程进行操作。代码如下：

```
/*****************************************************************************
* 名称: threadManage(int argc, char **argv)
* 功能: 自定义 FinSH 控制台命令
* 参数: argc 表示参数个数；argv 表示指向参数字符串指针的数组指针
*****************************************************************************/
#include <rtthread.h>
#include "apl_led.h"
int threadManage(int argc, char **argv)
{
    rt_err_t result = RT_EOK;
    if(led_thread->type == 0)                    //判断 LED 线程是否被删除，避免程序异常
    {
        rt_kprintf("led thread have been removed!\n");
        return 0;
    }
    if(!rt_strcmp(argv[1], "start"))
    {
        result = rt_thread_startup(led_thread);      //启动 LED 线程
        if(result == RT_EOK)
            rt_kprintf("led thread starting success!\n");
        else
            rt_kprintf("led thread failed to start, errCode:%d\n", result);
    } else if(!rt_strcmp(argv[1], "suspend")) {
        if(suspendFlag == 0)
            suspendFlag = 1;                          //置位挂起线程标志位
        if(result == RT_EOK)
            rt_kprintf("led thread suspend success!\n");
        else
            rt_kprintf("led thread suspend failed, errCode:%d\n", result);
    } else if(!rt_strcmp(argv[1], "resume")) {
        result = rt_thread_resume(led_thread);        //恢复线程
        if(result == RT_EOK)
            rt_kprintf("led thread resume success!\n");
        else
            rt_kprintf("led thread resume failed, errCode:%d\n", result);
    }
    else if(!rt_strcmp(argv[1], "delete"))
    {
        result = rt_thread_delete(led_thread);        //删除线程
```

```
            if(result == RT_EOK)
                rt_kprintf("led thread deletion success!\n");
            else
                rt_kprintf("led thread fail to delete, errCode:%d\n", result);
        }
        else
            rt_kprintf("Please input 'threadManage <start|suspend|resume|delete>'\n");
        return 0;
    }
    MSH_CMD_EXPORT(threadManage, thread Manage sample);        //导出自定义的 FinSH 控制台命令
```

2.1.3　开发步骤与验证

2.1.3.1　硬件部署

同 1.2.3.1 节。

2.1.3.2　工程调试

（1）将工程文件夹（05-Thread）复制到"RT-ThreadStudio/workspace"目录下。

（2）部署公共文件（02-软件资料/01-操作系统/rtt-common.zip）：将 rt-thread 文件夹复制到工程的根目录、将 zonesion/common 文件夹复制到工程的 zonesion 目录下。

（3）使用 RT-Thread Studio 导入工程。

完成工程的编译、下载后即可进行调试。

2.1.3.3　验证效果

（1）关闭 RT-Thread Studio，拔掉仿真器，按下 ZI-ARMEmbed 上的电源按键重新上电。

（2）在 MobaXterm 串口终端 FinSH 控制台中输入"threadManage start"命令，解析自定义 FinSH 控制台命令后，启动 LED 线程，4 个 LED 循环点亮（见图 2.5），同时 FinSH 控制台输出"led thread starting success!"信息。

```
 \ | /
- RT -     Thread Operating System
 / | \      4.1.0 build Oct 13 2022 17:55:30
 2006 - 2022 Copyright by RT-Thread team
msh >threadManage start
led thread starting success!
msh >
```

图 2.5　4 个 LED 循环点亮

（3）在 MobaXterm 串口终端 FinSH 控制台中输入"threadManage suspend"命令，解析自定义 FinSH 控制台命令后，挂起 LED 线程，4 个 LED 停止循环点亮，这时只有 1 个 LED 点亮，同时 FinSH 控制台输出"led thread suspend success!"信息。

```
 \ | /
- RT -     Thread Operating System
 / | \     4.1.0 build Oct 13 2022 17:55:30
 2006 - 2022 Copyright by RT-Thread team
msh >threadManage start
led thread starting success!
msh >threadManage suspend
led thread suspend success!
msh >
```

（4）在 MobaXterm 串口终端 FinSH 控制台中输入"threadManage resume"命令，解析自定义 FinSH 控制台命令后，恢复 LED 线程，4 个 LED 继续循环点亮，同时 FinSH 控制台输出"led thread resume success!"信息。

```
 \ | /
- RT -     Thread Operating System
 / | \     4.1.0 build Oct 13 2022 17:55:30
 2006 - 2022 Copyright by RT-Thread team
msh >threadManage start
led thread starting success!
msh >threadManage suspend
led thread suspend success!
msh >threadManage resume
led thread resume success!
msh >
```

（5）在 MobaXterm 串口终端 FinSH 控制台中输入"threadManage delete"命令，解析自定义 FinSH 控制台命令后，删除 LED 线程，4 个 LED 停止循环点亮，这时只有 1 个 LED 点亮，同时 FinSH 控制台输出"led thread delete success!"信息。使用"list_thread"命令可以看到此时 LED 线程已被删除。

```
 \ | /
- RT -     Thread Operating System
 / | \     4.1.0 build Oct 13 2022 17:55:30
 2006 - 2022 Copyright by RT-Thread team
msh >threadManage start
led thread starting success!
msh >threadManage suspend
led thread suspend success!
msh >threadManage resume
led thread resume success!
msh >threadManage delete
led thread delete success!
msh >
```

2.1.4 小结

本节介绍了线程管理的基本概念、工作原理和管理方式。通过本节的学习，读者可掌握线程管理接口的应用。

2.2 RT-Thread 定时器应用开发

定时器在嵌入式系统中是一种常见的机制，用于在一定的时间间隔内执行特定的任务。定时器通常用于处理周期性任务、轮询、事件调度等。在不同的环境中，定时器有不同的实现方式和应用场景。RT-Thread 提供了灵活且功能丰富的定时器机制，RT-Thread 的定时器可分为：

（1）软件定时器：RT-Thread 支持软件定时器，这是通过系统时钟的滴答定时器（Tick Timer）来实现的，允许用户在应用程序中创建基于时间的事件。

（2）硬件定时器：RT-Thread 还支持硬件定时器，硬件定时器通常是由微控制器上的硬件定时器或定时器模块实现的。

本节的要求如下：

- ➲ 了解硬件定时器的基本概念。
- ➲ 学习定时器的工作原理。
- ➲ 掌握定时器的管理方式。
- ➲ 掌握 RT-Thread 定时器的应用。

2.2.1 原理分析

2.2.1.1 定时器的基本概念

通俗地讲，定时器是指从指定的时刻开始，经过一定的时间后触发一个事件。定时器有硬件定时器和软件定时器之分。

（1）硬件定时器。硬件定时器是芯片本身提供的定时器，一般采用由外部晶振提供给芯片的输入时钟，芯片向软件模块提供一组配置寄存器，在到达设定的时间值后芯片的中断控制器产生时钟中断。硬件定时器的精度一般很高，可以达到纳秒级，并且采用中断触发方式。

（2）软件定时器。软件定时器是由操作系统提供的一类系统接口，它构建在硬件定时器的基础之上，使系统能够提供不受数目限制的定时器服务。

RT-Thread 提供了软件定时器，以系统时钟的时间长度为单位，即定时数值必须是系统时钟时间长度的整数倍。例如，如果系统时钟的时间长度是 10 ms，那么软件定时器只能是 10 ms、20 ms、100 ms 等的定时器，而不能为是 15 ms 的定时器。RT-Thread 的软件定时器也是基于系统时钟的，提供了基于系统时钟时间长度整数倍的定时能力，本节的系统时钟时间长度为 1 ms。

RT-Thread 的定时器提供两类定时器机制：第一类是单次触发定时器，这类定时器在启动后只会触发一次定时器事件，然后定时器自动停止；第二类是周期触发定时器，这类定时器会周期性地触发定时器事件，直到用户手动停止为止，否则将继续执行下去。

另外，根据超时函数执行时所处的上下文环境，RT-Thread 定时器的工作模式可分为 HARD_TIMER 模式与 SOFT_TIMER 模式，如图 2.6 所示。

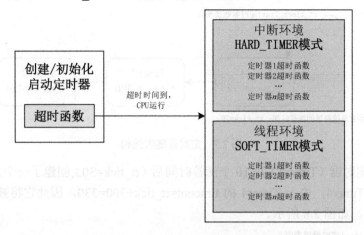

图 2.6 RT-Thread 定时器的工作模式

1）HARD_TIMER 模式

在 HARD_TIMER 模式下，定时器超时函数在中断上下文环境中执行，可以在初始化/创建定时器时使用参数 RT_TIMER_FLAG_HARD_TIMER 来指定定时器工作在该模式下。

RT-Thread 定时器的默认工作模式是 HARD_TIMER 模式，即定时器超时后，超时函数是在系统时钟中断的上下文环境中运行的。在中断上下文中的执行方式决定了定时器的超时函数不会调用任何会让当前上下文挂起的系统函数，也不能执行非常长的时间，否则会导致其他中断的响应时间加长或抢占其他线程执行的时间。

2）SOFT_TIMER 模式

SOFT_TIMER 模式是可配置的，用户既可以通过宏定义 RT_USING_TIMER_SOFT 决定定时器是否工作于该模式，也可以在初始化/创建定时器时使用参数 RT_TIMER_FLAG_SOFT_TIMER 来指定定时器工作于该模式。SOFT_TIMER 模式在启用后，系统会在初始化时创建一个 timer 线程，定时器的超时函数会在 timer 线程的上下文环境中执行。

2.2.1.2 定时器的工作原理

下面以一个例子来说明 RT-Thread 定时器的工作原理。RT-Thread 的定时器模块维护着两个重要的全局变量：

（1）当前系统经过的滴答时间 rt_tick：当硬件定时器中断触发时，该全局变量将加 1。

（2）定时器链表 rt_timer_list：系统创建并激活的定时器都会按照超时时间排序的方式插入 rt_timer_list 中。

定时器链表的结构如图 2.7 所示，系统当前时间为 20 个滴答时间，在当前系统中已经创建并启动了三个定时器，分别是定时时间为 50 个滴答时间的 Timer1、100 个滴答时间的 Timer2 和 500 个滴答时间的 Timer3，这三个定时器分别加上系统当前时间 rt_tick=20，从小到大排序链接在 rt_timer_list 中。

rt_tick 随着硬件定时器的中断触发次数的增加而增长（每触发一次硬件定时器中断，rt_tick 就会加 1），经过 50 个滴答时间后，rt_tick 从 20 增长到 70，与 Timer1 的 timeout 值相等，这时会触发与 Timer1 定时器相关联的超时函数，同时将 Timer1 从 rt_timer_list 中删除。

同理，经过 100 个滴答时间和 500 个滴答时间后，与 Timer2 和 Timer3 定时器相关联的超时函数会被触发，接着将 Time2 和 Timer3 从 rt_timer_list 中删除。

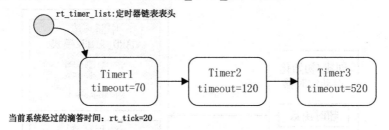

图 2.7   定时器链表结构

如果当前定时器（Timer1）在 10 个滴答时间后（rt_tick=30）创建了一个定时时间为 300 个滴答时间的 Timer4，由于 Timer4 的 timeout=rt_tick+300=330，因此它将被插入在 Timer2 和 Timer3 中间，如图 2.8 所示。

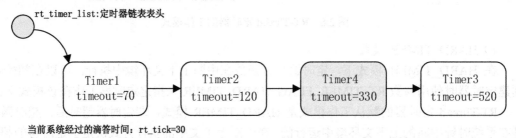

图 2.8   新创建的定时器在定时器链表中的位置

2.2.1.3   定时器的管理方式

前文介绍了 RT-Thread 定时器的工作原理，本节将深入介绍定时器的各个接口，帮助读者在代码层次上理解 RT-Thread 定时器的管理方式。

RT-Thread 在启动时需要初始化定时器管理系统。通过下面的函数可完成定时器管理系统的初始化：

```
/**********************************************************************************
 * 名称：rt_system_timer_init()
 * 功能：初始化定时器管理系统
 **********************************************************************************/
void rt_system_timer_init(void);
```

如果需要使用 SOFT_TIMER 模式，则在初始化定时器管理系统时应该调用下面的函数：

```
/**********************************************************************************
 * 名称：rt_system_timer_thread_init()
 * 功能：初始化软件定时器管理系统
 **********************************************************************************/
void rt_system_timer_thread_init(void);
```

定时器控制块中包含了与定时器相关的重要参数，这些参数在定时器的各种状态间起到纽带作用。定时器的相关操作如图 2.9 所示。所有的定时器都会在定时超时后从定时器链表中删除，而周期性定时器则会在再次启动时加入定时器链表，这与定时器参数设置相关。当操作系统时钟中断触发时，都会对已经超时的定时器状态参数进行修改。

图 2.9　定时器的相关操作

1）创建定时器

在需要动态创建一个定时器时，可使用下面的函数：

```
/**********************************************************************************
 * 名称：rt_timer_t rt_timer_create()
 * 功能：动态创建一个定时器
 * 参数：name 表示定时器的名称；void (timeout) (void parameter)表示定时器超时函数指针（当定时
器超时时，系统会调用这个函数）；parameter 表示定时器超时函数的入口参数（当定时器超时时，调用超
时回调函数会把这个参数作为入口参数传递给超时函数）；time 表示定时器的超时时间，单位是时钟节
拍；flag 表示定时器在创建时的参数，支持的值包括单次触发定时器、周期触发定时器、硬件定时器、软件定时
器等（可以用"或"关系取多个值）
 * 返回：RT_NULL 表示创建失败（通常会由于系统内存不够用而返回 RT_NULL）；定时器句柄表
示定时器创建成功
 **********************************************************************************/
 rt_timer_t rt_timer_create(const char*  name, void (*timeout)  (void* parameter), void*  parameter,
                  rt_tick_t   time, rt_uint8_t  flag);
```

调用上面的函数后，RT-Thread 会首先为定时器控制块分配动态内存，然后对该定时器控制块进行基本的初始化。rtdef.h 中定义了一些与定时器相关的宏，例如：

```
#define RT_TIMER_FLAG_ONE_SHOT        0x0                  //单次触发定时器
#define RT_TIMER_FLAG_PERIODIC        0x2                  //周期触发定时器
#define RT_TIMER_FLAG_HARD_TIMER      0x0                  //硬件定时器
#define RT_TIMER_FLAG_SOFT_TIMER      0x4                  //软件定时器
```

当指定的 flag 为 RT_TIMER_FLAG_HARD_TIMER 时，如果定时器超时，则定时器的回调函数将在时钟中断服务程序上下文中被调用；当指定的 flag 为 RT_TIMER_FLAG_SOFT_TIMER 时，如果定时器超时，则定时器的回调函数将在系统时钟 timer 线程的上下文中被调用。

2）删除定时器

当系统不再使用定时器时，可使用下面的函数将其删除：

```
/**********************************************************************************
 * 名称：rt_timer_delete()
 * 功能：删除一个动态定时器
 * 参数：timer 表示定时器句柄，指向要删除的定时器
```

```
 * 返回：RT_EOK 表示删除成功（如果参数 timer 是 RT_NULL，将会导致 ASSERT 错误）
 *****************************************************************************/
 rt_err_t rt_timer_delete(rt_timer_t timer);
```

3）启动定时器

定时器在创建或者初始化后，并不会被立即启动，必须调用启动定时器的函数后定时器才开始工作。启动定时器的函数如下：

```
/*****************************************************************************
 * 名称：rt_timer_start()
 * 功能：启动一个定时器
 * 参数：timer 表示定时器句柄，指向要启动的定时器控制块
 * 返回：RT_EOK 表示启动成功
 *****************************************************************************/
 rt_err_t rt_timer_start(rt_timer_t   timer);
```

调用定时器启动函数后，定时器的状态将更改为激活状态（RT_TIMER_FLAG_ACTIVATED），并按照超时顺序插入 rt_timer_list 中。

4）停止定时器

在启动定时器后，若想停止它，可以使用下面的函数：

```
/*****************************************************************************
 * 名称：rt_timer_stop()
 * 功能：停止一个定时器
 * 参数：timer 表示定时器句柄，指向要停止的定时器控制块
 * 返回：RT_EOK 表示成功停止定时器；-RT_ERROR 表示定时器已经处于停止状态
 *****************************************************************************/
 rt_err_t rt_timer_stop(rt_timer_t   timer);
```

调用定时器停止函数后，定时器的状态将更改为停止状态，并从 rt_timer_list 中脱离出来不参与定时器超时检查。当某个定时器超时时，也可以调用这个函数停止该定时器本身。

5）控制定时器

除了上述提供的一些函数，RT-Thread 还提供了定时器控制函数，以获取或设置更多定时器的信息。控制定时器的函数如下：

```
/*****************************************************************************
 * 名称：rt_timer_control()
 * 功能：设置定时器
 * 参数：timer 表示定时器句柄，指向要操作的定时器控制块
 * 参数：cmd 表示用于控制定时器的命令，当前支持 4 条命令，分别是设置定时时间、查看定时时
间、设置单次触发定时器、设置周期触发定时器；arg 与 cmd 表示对应的控制命令的参数，如 cmd 为设定
的超时时间，就可以通过 arg 设定超时时间
 * 返回：RT_EOK 表示修改成功
 *****************************************************************************/
 rt_err_t rt_timer_control(rt_timer_t   timer, rt_uint8_t   cmd, void*   arg);
```

控制定时器函数可根据命令类型来查看或改变定时器的设置。

2.2.2　开发设计与实践

2.2.2.1　硬件设计

本节的硬件设计同 1.5.2.1 节。

2.2.2.2　软件设计

软件设计流程如图 2.10 所示。

（1）创建定时器线程，并且在 main 函数中完成线程的初始化工作。

（2）编写线程入口函数，在线程入口函数中完成 LED 引脚的初始化工作。

（3）在 MobaXterm 串口终端 FinSH 控制台中输入"timerManage start"命令，4 个 LED 循环点亮，同时在 FinSH 控制台中输出"led timer starting success!"信息；当输入"timerManage control 100"命令时，4 个 LED 循环点亮的速度加快，同时在 FinSH 控制台中输出"led timer control success!"信息；当输入"timerManage stop"命令时，4 个 LED 停止循环点亮，只有 1 个 LED 点亮，同时在 FinSH 控制台中输出"led timer stop success!"信息；当输入"timerManage delete"命令时，4 个 LED 循环点亮停止，同时在 FinSH 控制台中输出"led timer deletion success!"信息。

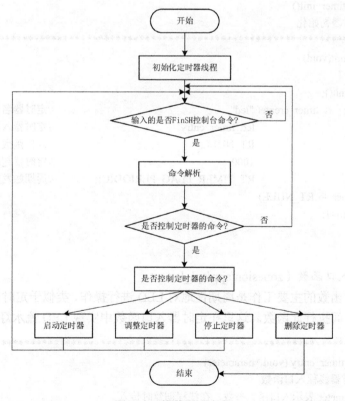

图 2.10　软件设计流程

2.2.2.3　功能设计与核心代码设计

通过原理学习可知，要实现定时器的管理，就需要使用 RT-Thread 提供的定时器管理接口进行定时器的初始化。定时器的管理使用的是 FinSH 控制台命令，需要实现对应的 FinSH

控制台命令，并将已实现命令添加到 FinSH 控制台命令列表中。

1）主函数（zonesion/app/main.c）

主函数的主要工作是调用 led_timer_init() 函数，完成引导工作。代码如下：

```
/*******************************************************************************
 * 名称：main()
 * 功能：调用 led_timer_init() 函数，完成引导工作
 *******************************************************************************/
#include "apl_led.h"
int main(void)
{
    led_timer_init();                                    //定时器线程初始化
    return 0;
}
```

2）定时器初始化函数（zonesion/app/apl_led.c）

定时器初始化函数的主要工作是实现定时器的创建，包括定义定时器名称和入口函数，设置入口函数参数、定时器超时时间、定时器类型。代码如下：

```
/*******************************************************************************
 * 名称：led_timer_init()
 * 功能：定时器初始化
 *******************************************************************************/
int led_timer_init(void)
{
    led_pin_init();
    led_timer = rt_timer_create( "led",                      //定时器名称
                            led_timer_entry,                 //定时器入口函数
                            RT_NULL,                         //入口函数参数
                            1000,                            //定时器超时时间
                            RT_TIMER_FLAG_PERIODIC);         //周期触发定时器
    if(led_timer == RT_NULL)
        return -1;
    return 0;
}
```

3）定时器入口函数（zonesion/app/apl_led.c）

定时器入口函数的主要工作是周期性地对 LED 进行操作，类似于定时器中断服务程序，需要在规定的时间执行此函数。这里在定时器入口函数中实现 LED 流水灯。代码如下：

```
/*******************************************************************************
 * 名称：led_timer_entry (void *parameter)
 * 功能：定时器线程入口函数
 * 参数：parameter 表示入口函数参数，在线程创建时传入
 *******************************************************************************/
static void led_timer_entry(void *parameter)
{
    (void)parameter;
    static unsigned char ledState = 0x01;
```

```
        led_ctrl(ledState);
        if(ledState < 0x08)
            ledState <<= 1;
        else
            ledState = 0x01;
}
```

4）自定义 FinSH 控制台命令函数（zonesion/app/finsh_cmd/finsh_ex.c）

自定义 FinSH 控制台命令函数的主要工作是实现定时器管理的命令，包括定时器的启动、控制、停止及删除，可以在 FinSH 控制台中使用相关的命令对定时器进行操作。代码如下：

```
/**************************************************************************
* 名称：timerManage (int argc, char **argv)
* 功能：自定义 FinSH 控制台命令
* 参数：argc 表示参数个数；argv 表示指向参数字符串指针的数组指针
**************************************************************************/
#include <rtthread.h>
#include <stdlib.h>
#include "apl_led.h"
int timerManage(int argc, char **argv)
{
    rt_err_t result = RT_EOK;
    if(led_timer->parent.type == 0)                  //判断定时器线程是否被删除，避免程序异常
    {
        rt_kprintf("led timer have been removed!\n");
        return 0;
    }
    if(!rt_strcmp(argv[1], "start"))
    {
        result = rt_timer_start(led_timer);          //启动定时器
        if(result == RT_EOK)
            rt_kprintf("led timer starting success!\n");
        else
            rt_kprintf("led timer failed to start, errCode:%d\n", result);
    } else if(!rt_strcmp(argv[1], "control")) {
        if(argc < 3)
        {
            rt_kprintf("Please input timeout parameter!\n");
            return 0;
        }
        int timeOut = atoi(argv[2]);
        if(timeOut < 100)
        {
            rt_kprintf("The timeout needs to be greater than 100!\n");
            return 0;
        }
        result = rt_timer_control(led_timer, RT_TIMER_CTRL_SET_TIME, &timeOut);//修改定时器超时时间
        if(result == RT_EOK)
```

```
                    rt_kprintf("led timer control success!\n");
            else
                    rt_kprintf("led timer control failed, errCode:%d\n", result);
        } else if(!rt_strcmp(argv[1], "stop")) {
            result = rt_timer_stop(led_timer);                              //停止定时器
            if(result == RT_EOK)
                    rt_kprintf("led timer stop success!\n");
            else
                    rt_kprintf("led timer stop failed, errCode:%d\n", result);
        } else if(!rt_strcmp(argv[1], "delete")) {
            result = rt_timer_delete(led_timer);                           //删除定时器
            if(result == RT_EOK)
                    rt_kprintf("led timer deletion success!\n");
            else
                    rt_kprintf("led timer fail to delete, errCode:%d\n", result);
        }
        else
            rt_kprintf("Please input 'timerManage <start|control|stop|delete>'\n");
        return 0;
    }
MSH_CMD_EXPORT(timerManage, timer Manage sample);
```

2.2.3　开发步骤与验证

2.2.3.1　硬件部署

同 1.2.3.1 节。

2.2.3.2　工程调试

（1）将本项目的工程（06-Timer）文件夹复制到 RT-ThreadStudio\workspace 目录下。

（2）其余同 2.1.3.2 节。

2.2.3.3　验证效果

（1）关闭 RT-Thread Studio，拔掉仿真器，按下 ZI-ARMEmbed 上的电源按键重新上电。

（2）在 MobaXterm 串口终端 FinSH 控制台中输入"timerManage start"命令，4 个 LED 循环点亮，点亮的时间间隔是 1 s（见图 2.11），同时在 FinSH 控制台中输出"led timer starting success!"信息。

```
 \ | /
- RT -     Thread Operating System
 / | \      4.1.0 build Oct 14 2022 09:04:02
 2006 - 2022 Copyright by RT-Thread team
msh >timerManage start
led timer starting success!
msh >
```

图 2.11　4 个 LED 循环点亮，点亮的时间间隔是 1 s

（3）在 MobaXterm 串口终端 FinSH 控制台中输入 "timerManage control 100" 命令，将定时器设置为 100 ms 的定时器，4 个 LED 循环点亮，循环的速度将变快，同时在 FinSH 控制台中输出 "led timer control success!" 信息。

```
 \ | /
- RT -      Thread Operating System
 / | \      4.1.0 build Oct 14 2022 09:04:02
 2006 - 2022 Copyright by RT-Thread team
msh >timerManage start
led timer starting success!
msh >timerManage control 100
led timer control success!
msh >
```

（4）在 MobaXterm 串口终端 FinSH 控制台中输入 "timerManage stop" 命令，将停止定时器，4 个 LED 停止循环点亮，同时在 FinSH 控制台中输出 "led timer stop success!" 信息。

```
 \ | /
- RT -      Thread Operating System
 / | \      4.1.0 build Oct 14 2022 09:04:02
 2006 - 2022 Copyright by RT-Thread team
msh >timerManage start
led timer starting success!
msh >timerManage control 100
led timer control success!
msh >timerManage stop
led timer stop success!
msh >
```

（4）在 MobaXterm 串口终端 FinSH 控制台中输入 "timerManage delete" 命令，将删除定时器，在 FinSH 控制台中输出 "led timer delete success!" 信息，使用 "list_thread" 命令可以看到此时定时器已被删除。

```
 \ | /
- RT -      Thread Operating System
 / | \      4.1.0 build Oct 14 2022 09:04:02
 2006 - 2022 Copyright by RT-Thread team
msh >timerManage start
led timer starting success!
```

```
msh >timerManage control 100
led timer control success!
msh >timerManage stop
led timer stop success!
msh >timerManage delete
led timer delete success!
msh >list_thread
thread    pri   status      sp          stack size max used left tick   error
--------  ---   -------  ----------  ---------- ------  ---------- ---
tshell     20   running  0x000000cc 0x00001000     14%    0x00000006 000
tidle0     31   ready    0x00000084 0x00000100     61%    0x00000003 000
timer       4   suspend  0x0000007c 0x00000200     24%    0x00000009 000
msh >
```

2.2.4　小结

本节主要介绍定时器的基本概念、工作原理和管理方式。通过本节的学习，读者可以掌握 RT-Thread 定时器的应用。

2.3 RT-Thread 信号量应用开发

在操作系统和并发编程中，信号量是一个重要的同步机制，被广泛用于解决临界区问题（Critical Section Problem）和进程间通信（Inter-Process Communication，IPC）等场景。

信号量（Semaphore）是一种用于控制多个进程或线程对共享资源进行访问的机制，它是由荷兰计算机科学家 Edsger Dijkstra 于 1965 年提出的。信号量的一个主要用途是在多进程或多线程环境中协调对共享资源的访问。通过使用信号量，可以确保在任何给定时刻，只有一个进程或线程能够访问共享资源，从而避免数据竞争和不一致性。

本节的要求如下：

➾ 了解信号量的基本概念。
➾ 学习信号量的工作原理。
➾ 掌握信号量的管理方式。
➾ 掌握 RT-Thread 信号量的应用。

2.3.1　原理分析

2.3.1.1　信号量

信号量通常是一个整型变量，用于表示某种资源的数量。这个整数值可以是任意非负数。信号量的操作包括两个原子操作：增加（通常称为 V 操作）和减少（通常称为 P 操作）。这两个操作是原子的，不会被中断。

P（Produce）操作：也称为 down 操作，用于申请资源。如果信号量值大于零，就减 1 并继续执行；否则，阻塞等待。

V（Vacate）操作：也称为 up 操作，用于释放资源。如果有其他进程等待该资源，就唤醒一个等待的进程；否则，信号量加 1。

信号量同步的原理借鉴了操作系统中所用到的 PV 原语：一次 P 操作使信号量 sem 减 1，一次 V 操作使 sem 加 1。进程或线程根据信号量值来判断自己是否具有对公共资源进行访问的权限。当 sem 的值大于或等于 0 时，该进程或线程具有对公共资源进行访问的权限；否则，当 sem 的值小于 0 时，该进程或线程将阻塞，直到 sem 的值大于或等于 0 时为止。信号量的操作如图 2.12 所示。

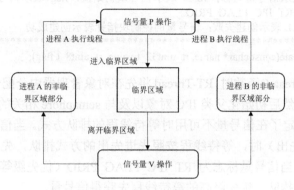

图 2.12　信号量的操作

2.3.1.2　信号量的工作原理

信号量是一种轻型的用于解决线程间同步问题的内核对象，线程可以获取或释放它，从而达到同步或互斥的目的。信号量的工作原理如图 2.13 所示，每个信号量对象都有一个信号量值和一个线程等待队列，信号量值对应着信号量对象的实例数量、资源数量。

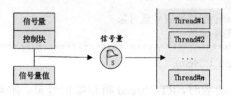

图 2.13　信号量的工作原理

假如信号量值为 5，则表示共有 5 个信号量实例（资源）可以被使用，当信号量实例数量为零时，再申请该信号量的线程就会被挂起在该信号量的线程等待队列上，等待可用的信号量实例（资源）。

2.3.1.3　信号量的管理方式

信号量控制块包括了与信号量相关的重要参数，这些参数在信号量各种状态间起到纽带的作用。信号量控制块的相关函数如图 2.14 所示。

图 2.14　信号量控制块的相关函数

1）创建信号量

在创建一个信号量时，RT-Thread 首先创建一个信号量控制块，然后对该控制块进行基本的初始化工作。使用下面的函数可创建信号量：

```
/******************************************************************************
 * 名称：rt_sem_create()
 * 功能：信号量创建函数
 * 参数：name 表示信号量名称；value 表示信号量初始值；flag 表示信号量标志，其取值可以是
RT_IPC_FLAG_FIFO 或 RT_IPC_FLAG_PRIO
 * 返回：RT_NULL 表示创建失败；信号量的控制块指针表示创建成功
 ******************************************************************************/
rt_sem_t rt_sem_create(const char* name, rt_uint32_t value, rt_uint8_t flag);
```

当调用 rt_sem_create()函数时，RT-Thread 将先在对象管理器中分配一个 semaphore 对象，并初始化这个对象，然后初始化父类 IPC 对象以及与 semaphore 相关的部分。在信号量的参数中，信号量标志决定了在信号量不可用时等待线程的排队方式。当信号量标志为 RT_IPC_FLAG_FIFO（先进先出）时，等待线程按照先进先出的方式排队，先进入线程等列队列的线程先获得信号量；当信号量标志为 RT_IPC_FLAG_PRIO（优先级等待）时，等待线程将按照优先级高低进行排队，优先级高的等待线程先获得信号量。

2）删除信号量

当系统不再使用信号量时，可通过删除信号量来释放系统资源。通过下面的函数可删除信号量（适用于动态创建的信号量）：

```
/******************************************************************************
 * 名称：rt_sem_delete()
 * 功能：信号量删除函数
 * 参数：sem rt_sem_create()创建的信号量对象
 * 返回：RT_EOK 表示删除成功
 ******************************************************************************/
rt_err_t rt_sem_delete(rt_sem_t  sem);
```

在调用 rt_sem_delete()函数时，RT-Thread 将删除信号量。如果线程正在等待被删除的信号量，那么删除操作会先唤醒等待该信号量的线程（等待线程的返回值是"-RT_ERROR"），然后释放信号量的内存资源。

3）获取信号量

线程通过获取信号量可获得信号量的实例（资源），当信号量值大于零时，线程将获得信号量，并且信号量值会减 1。通过下面的函数可获取信号量：

```
/******************************************************************************
 * 名称：rt_sem_take()
 * 功能：获取信号量
 * 参数：sem 表示信号量对象句柄；time 表示等待时间，单位是操作系统时钟节拍
 * 返回：RT_EOK 表示成功获得信号量；-RT_ETIMEOUT 表示超时依然未获得信号量；-RT_ERROR
表示其他错误
 ******************************************************************************/
rt_err_t rt_sem_take (rt_sem_t  sem, rt_int32_t  time);
```

在调用 rt_sem_take()函数时，如果信号量值等于零，则说明当前信号量的实例（资源）不可用，申请该信号量的线程将根据参数 time 的情况选择直接返回、挂起等待一段时间或永久等待，直到其他线程或中断释放该信号量为止。如果在参数 time 指定的时间内依然得

不到信号量，线程将超时返回，返回值是"-RT_ETIMEOUT"。

　　4）释放信号量

释放信号量可以唤醒挂起在该信号量上的线程。通过下面的函数可释放信号量：

```
/*****************************************************************************
 * 名称：rt_sem_release()
 * 功能：释放信号量
 * 参数：sem 表示信号量对象句柄
 * 返回：RT_EOK 表示成功释放信号量
 *****************************************************************************/
rt_err_t rt_sem_release(rt_sem_t    sem);
```

　　例如，当信号量值等于零，并且有线程等待这个信号量时，释放信号量将唤醒在该信号量的线程等待队列中的第一个线程，由第一个线程获取信号量；否则将信号量值加 1。

2.3.2　开发设计与实践

2.3.2.1　硬件设计

本节的硬件设计同 1.5.2.1 节。本节首先通过 MobaXterm 串口终端 FinSH 控制台命令释放信号量，然后在 LED 线程中获取这个信号量后，实现一次 LED 流水灯。

2.3.2.2　软件设计

软件设计流程如图 2.15 所示。

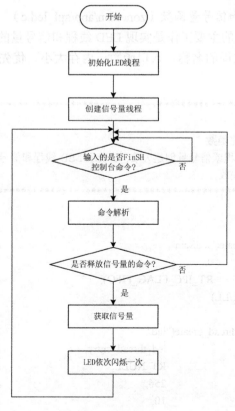

图 2.15　软件设计流程

（1）创建信号量线程和 LED 线程，并且在 main 函数中完成线程的初始化。

（2）编写 LED 线程的入口函数，在其中添加获取信号量的等待条件。

（3）在 MobaXterm 串口终端 FinSH 控制台中输入"sem release"命令，实现一次 LED 流水灯的效果，同时在 FinSH 控制台中输出"The led thread gets a semaphore and uses it"。

2.3.2.3　功能设计与核心代码设计

通过原理学习可知，要实现信号量的管理，就需要先创建一个信号量，然后自定义 FinSH 控制台命令，通过命令释放信号量，最后在 LED 线程的入口函数中获取信号量，实现一次 LED 流水灯效果，并且流水灯的实现和释放信号量的次数相同。

1）主函数（zonesion/app/main.c）

主函数的主要工作是调用 led_thread_init()函数，完成引导工作。代码如下：

```
/*****************************************************************************
 * 名称：main()
 * 功能：调用 led_thread_init()函数，完成引导工作
/*****************************************************************************
#include "apl_led.h"
int main(void)
{
    led_thread_init();                                      //LED 线程初始化
    return 0;
}
```

2）LED 线程初始化和信号量函数（zonesion/app/apl_led.c）

LED 线程初始化函数的主要工作是实现 LED 线程和信号量的创建，包含信号量的名称、初始个数、模式，LED 线程的名称、入口函数、内存大小、优先级、时间片等参数。代码如下：

```
/*****************************************************************************
 * 名称：led_thread_init()
 * 功能：LED 线程初始化函数
 * 返回：-1 表示 LED 线程或信号量创建失败；0 表示 LED 线程和信号量创建并启动成功；其他值表
示 LED 线程创建成功但启动失败
 *****************************************************************************/
int led_thread_init(void)
{
    led_sem = rt_sem_create("ledSem",                       //信号量名称
                        0,                                  //信号量初始个数
                        RT_IPC_FLAG_FIFO);                  //先进先出模式
    if(led_sem == RT_NULL)
        return -1;
    led_thread = rt_thread_create("led",                    //LED 线程名称
                        led_thread_entry,                   //LED 线程入口函数
                        RT_NULL,                            //LED 线程入口函数参数
                        256,                                //LED 线程内存大小
                        10,                                 //LED 线程优先级
                        20);                                //LED 线程时间片
```

```
        if(led_thread == RT_NULL)
            return -1;
        return rt_thread_startup(led_thread);                        //启动 LED 线程
}
```

3）线程入口函数（zonesion/app/apl_led.c）

线程入口函数的主要工作是完成 LED 引脚的初始化，并且在死循环中等待信号量，当 LED 线程获取到信号量后，实现一次 LED 流水灯效果。代码如下：

```
/********************************************************************************
* 名称：led_thread_entry()
* 功能：LED 线程入口函数
* 参数：*parameter 入口函数参数
********************************************************************************/
static void led_thread_entry(void *parameter)
{
    (void)parameter;
    rt_err_t result = RT_EOK;
    led_pin_init();                                                  //LED 引脚初始化
    while(1)
    {
        result = rt_sem_take(led_sem, RT_WAITING_FOREVER);           //一直等待信号量
        if(result == RT_EOK)
        {
            rt_kprintf("The led thread gets a semaphore and uses it!\n");
            for(unsigned char ledState=0x01; ledState<=0x10; ledState<<=1)
            {
                led_ctrl(ledState);                                  //控制 LED 的亮灭
                rt_thread_mdelay(500);                               //延时 500 ms
            }
        }
    }
}
```

4）自定义 FinSH 控制台命令函数（zonesion/app/finsh_cmd/finsh_ex.c）

自定义 FinSH 控制台命令函数的主要工作是释放信号量。代码如下：

```
/********************************************************************************
* 名称：sem (int argc, char **argv)
* 功能：信号量释放命令
* 参数：argc 表示参数个数；argv 表示指向参数字符串指针的数组指针
********************************************************************************/
#include <rtthread.h>
#include "apl_led.h"
int sem(int argc, char **argv)
{
    rt_err_t result = RT_EOK;
    if(!rt_strcmp(argv[1], "release"))
    {
```

```
            result = rt_sem_release(led_sem);                    //释放一个信号量
            if(result == RT_EOK)
                rt_kprintf("led semaphore release success!\n");
            else
                rt_kprintf("led semaphore release fail, errCode:%d\n", result);
        }
        else
            rt_kprintf("Please input 'sem <release>'\n");
        return 0;
    }
    MSH_CMD_EXPORT(sem, semaphore sample);
```

2.3.3　开发步骤与验证

2.3.3.1　硬件部署

同 1.2.3.1 节。

2.3.3.1　工程调试

（1）将本项目的工程（07-Semaphore）文件夹复制到 RT-ThreadStudio\workspace 目录下。

（2）其余同 2.1.3.2 节。

2.3.3.2　验证效果

（1）关闭 RT-Thread Studio，拔掉仿真器，按下 ZI-ARMEmbed 上的电源按键重新上电。

（2）在 MobaXterm 串口终端 FinSH 控制台中输入 "sem release" 命令，解析自定义 FinSH 控制台命令后，释放信号量，LED 线程获取到信号量后实现一次 LED 流水灯效果（见图 2.16），同时在 FinSH 控制台中输出 "The led thread gets a semaphore and uses it!" 信息。

```
 \ | /
- RT -     Thread Operating System
 / | \     4.1.0 build Oct 14 2022 09:15:13
 2006 - 2022 Copyright by RT-Thread team
msh >sem release
The led thread gets a semaphore and uses it!
led semaphore release success!
msh >
```

图 2.16　实现一次 LED 流水灯效果

2.3.4　小结

本节主要介绍信号量的基本概念、工作原理和管理方式。通过本节的学习，读者可掌握 RT-Thread 信号量的应用。

2.4 RT-Thread 互斥量应用开发

互斥量（Mutual Exclusion，Mutex）是一种用于确保在任意给定时刻只有一个线程能够访问共享资源的同步机制，它是多线程编程中常用的同步原语。互斥量提供了一种机制，可以防止多个线程同时访问共享资源，从而避免数据竞争和不一致性。一个线程在进入临界区（访问共享资源的代码段）之前必须先获得互斥量的锁，执行完临界区代码后再释放互斥量的锁。其他线程在试图获取这个互斥量的锁时会被阻塞，直到锁被释放为止。

本节的要求如下：
- 了解互斥量的基本概念。
- 学习互斥量的工作原理。
- 掌握互斥量的管理方式。
- 掌握 RT-Thread 互斥量的应用。

2.4.1　原理分析

2.4.1.1　互斥量

互斥量又称为相互排斥的信号量，是一种特殊的二值信号量。互斥量常用于保护共享资源，如在多线程环境中对数据结构的访问。使用互斥量可以确保在任何时刻只有一个线程能够修改共享资源，从而避免数据竞争和不一致性的问题。

互斥量类似于只有一个车位的停车场：当有一辆车进入时，将停车场大门锁住，其他车辆在外面等候；当里面的车出来时，将停车场大门打开，下一辆车才可以进入。

2.4.1.2　互斥量的工作原理

互斥量和信号量不同的是：持有互斥量的线程拥有互斥量的所有权，互斥量支持递归访问且能防止线程优先级翻转；互斥量只能由持有线程释放，而信号量则可以由其他线程释放。

互斥量的状态只有两种，开锁或闭锁（两种状态值）。互斥量的基本操作包括：

Lock（闭锁）：当一个线程尝试获取互斥量的锁时，如果互斥量当前未被锁定，该线程将获取锁并继续执行；否则，线程将被阻塞，直到锁被释放为止。

Unlock（开锁）：当一个线程释放互斥量的锁时，如果其他线程正在等待互斥量的锁，其中一个线程将被唤醒，并成功获取互斥量的锁。

当某个线程持有互斥量时，互斥量将处于闭锁状态，由该线程获得互斥量的所有权。当这个线程释放互斥量时，将对互斥量进行开锁，失去互斥量的所有权。当一个线程持有互斥量时，其他线程将不能够对互斥量进行开锁或持有它，持有互斥量的线程也能够再次获得这个互斥量而不被挂起。互斥量的工作原理如图 2.17 所示。

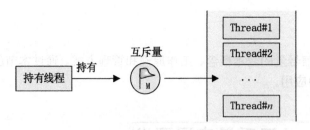

图 2.17   互斥量的工作原理

使用信号量会导致的一个潜在问题是线程优先级翻转。所谓优先级翻转，是指当一个高优先级线程试图通过信号量机制访问共享资源时，如果该信号量已被一低优先级线程持有，则这个低优先级线程在运行过程中可能又被其他的一些中等优先级的线程抢占，因此造成高优先级线程被许多具有较低优先级的线程阻塞，实时性难以得到保证。有三个线程 A、B、C，A 的优先级高于 B 的优先级，B 的优先级高于 C 的优先级，线程 A、B 处于挂起状态，等待某一事件触发，线程 C 正在运行，此时线程 C 开始使用共享资源 M。在 C 使用共享资源 M 的过程中，当线程 A 等待的事件触发时，线程 A 转为就绪态，因为它的优先级比线程 C 的优先级高，所以立即执行线程 A。但当线程 A 使用共享资源 M 时，由于共享资源 M 正在被线程 C 使用，因此线程 A 被挂起，切换到线程 C 运行。如果此时线程 B 等待的事件触发，则线程 B 转为就绪态。由于线程 B 的优先级比线程 C 高，因此线程 B 开始运行，直到其运行完毕，线程 C 才开始运行。只有当线程 C 释放共享资源 M 后，线程 A 才会运行。在这种情况下，优先级发生了翻转：线程 B 先于线程 A 运行，如图 2.18 所示。这样便不能保证高优先级线程的响应时间。

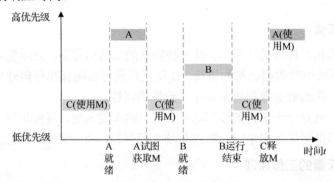

图 2.18   优先级翻转

在 RT-Thread 中，互斥量可以解决优先级翻转问题，解决的方法是使用优先级继承算法。优先级继承是指在线程 A 尝试获取共享资源 M 而被挂起的期间，将线程 C 的优先级提升到线程 A 的优先级，从而解决优先级翻转引起的问题，这样能够防止线程 C（间接地防止线程 A）被线程 B 抢占，如图 2.19 所示。优先级继承算法是指提高某个使用共享资源的线程优先级，使之与所有等待该共享资源的线程的最高优先级相等，然后执行使用共享资源的线程，当这个线程释放该共享资源时，将其优先级重新设置为原来的优先级。继承优先级的线程可避免共享资源被中间优先级的线程抢占。

注意：在获得互斥量后，请尽快释放互斥量，并且在持有互斥量的过程中，不得更改有互斥量的线程优先级。

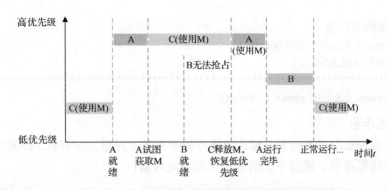

图 2.19　使用优先级继承算法解决优先级翻转问题

2.4.1.3　互斥量的管理方式

互斥量控制块包括了与互斥量相关的重要参数，这些参数在互斥量的实现中起到了纽带的作用。互斥量控制块的相关函数如图 2.20 所示。

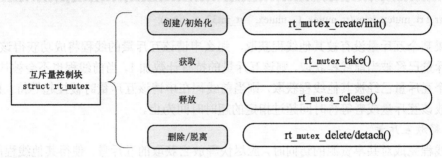

图 2.20　互斥量控制块的相关函数

1）创建互斥量

在创建一个互斥量时，RT-Thread 首先会创建一个互斥量控制块，然后对互斥量控制块进行初始化。通过下面的函数可创建互斥量：

```
/*********************************************************************
 * 名称：rt_mutex_create()
 * 功能：创建互斥量
 * 参数：name 表示互斥量的名称；flag 参数已经作废，无论用户选择 RT_IPC_FLAG_PRIO 还是
RT_IPC_FLAG_FIFO，RT-Thread 均按照 RT_IPC_FLAG_PRIO 处理
 * 返回：互斥量句柄表示创建成功；RT_NULL 表示创建失败
 *********************************************************************/
rt_mutex_t rt_mutex_create (const char*   name,   rt_uint8_t   flag);
```

在通过 rt_mutex_create() 函数创建一个互斥量时，互斥量的名称由参数 name 指定。当调用这个函数时，RT-Thread 先在对象管理器中分配一个 mutex 对象，并初始化这个对象，然后初始化父类 IPC 对象以及与 mutex 对象相关的部分。

2）删除互斥量

当不再使用互斥量时，可删除互斥量，从而释放资源。通过下面的函数可删除互斥量（适用于动态创建的互斥量）：

```
/*********************************************************************
 * 名称：rt_mutex_delete()
```

```
 * 功能：删除互斥量
 * 参数：mutex 表示互斥量对象句柄
 * 返回：RT_EOK 删除成功
 ***********************************************************************/
rt_err_t rt_mutex_delete (rt_mutex_t    mutex);
```

3）获取互斥量

如果某个线程获取了互斥量，那么该线程就拥有该互斥量的所有权，即某个时刻该互斥量只能被一个线程持有。通过下面的函数可获取互斥量：

```
/***********************************************************************
 * 名称：rt_mutex_take()
 * 功能：获取互斥量
 * 参数：mutex 表示互斥量对象句柄；time 表示等待的时间
 * 返回：RT_EOK 表示成功获得互斥量；-RT_ETIMEOUT 表示超时；-RT_ERROR 表示获取失败
 ***********************************************************************/
rt_err_t rt_mutex_take (rt_mutex_t    mutex,    rt_int32_t    time);
```

如果某个互斥量没有被其他线程获取，那么申请该互斥量的线程将成功获得该互斥量。如果互斥量已经被当前线程获取，则该互斥量的持有计数加 1，当前线程也不会被挂起等待。如果某个互斥量已经被其他线程获取，则当前线程在申请该互斥量时被挂起等待，直到其他线程释放该互斥量或者等待时间超过指定的超时时间为止。

4）释放互斥量

当线程完成对共享资源的访问时，应尽快释放它获取的互斥量，使得其他线程能及时获取互斥量。通过下面的函数可释放互斥量：

```
/***********************************************************************
 * 名称：rt_mutex_release()
 * 功能：释放互斥量
 * 参数：mutex 表示互斥量对象句柄
 ***********************************************************************/
rt_err_t rt_mutex_release(rt_mutex_t    mutex);
```

2.4.2　开发设计与实践

2.4.2.1　软件设计

软件设计流程如图 2.21 所示。软件设计上，首先创建 2 个线程，实现 2 个线程通过互斥量互锁。在创建 2 个线程时，线程 1 的优先级低于线程 2，线程 1 通过获取互斥量和不获取互斥量产生不同的结果。另外，如果线程 1 先获取互斥量的话，那么线程 1 此时会提升到线程 2 同等优先级。程序设计流程如下：

（1）创建 2 个线程（线程 1 的优先级低于线程 2）和 1 个互斥量，并且在 main 函数中完成线程的初始化。

（2）编写两个线程的入口函数，线程 1 通过是否被注释掉相关语句来控制是否获取互斥量，线程 2 用于显示线程 1 在执行时是否被抢占。

（3）在线程 1 获取互斥量时，MobaXterm 串口终端 FinSH 控制台持续输出"mutex

protect ,num1 = num2 is 2"并依次递增；如果将线程 thread1_entry 中的获取、释放互斥量的语句注释掉，则 MobaXterm 串口终端 FinSH 控制台输出"no protect , num1 = 1, num2 = 0"。

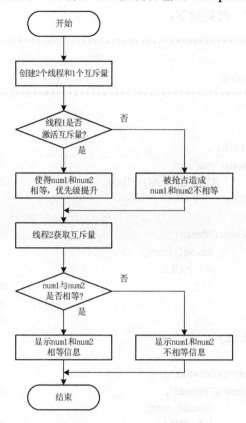

图 2.21　软件设计流程

在使用互斥量时，互斥量可以临时提升线程的优先级。如果线程 1 获取互斥量，则在线程 2 等待获取互斥量时，线程 1 可以将其优先级提升到和线程 2 同等的优先级。

2.4.2.2　功能设计与核心代码设计

通过原理学习可知，要掌握互斥量的使用，可以通过创建 2 个线程，它们的线程优先级是不同的，但通过互斥量可以使优先级较低的线程获取与高优先级线程同等优先级的权利。

1）主函数（zonesion/app/main.c）

主函数的主要工作是调用 thread_init()函数，完成引导工作。代码如下：

```
/***************************************************************************
 * 名称：main()
 * 功能：调用 thread_init()函数，完成引导工作
 /***************************************************************************
#include "mutex_sample.h"
int main(void)
{
    thread_init();                                    //互斥量线程初始化
    return 0;
}
```

2）互斥量线程初始化函数（zonesion/app/mutex_sample.c）

互斥量线程初始化函数的主要工作是创建 2 个线程和 1 个互斥量，包括定义 2 个线程的入口函数、互斥量名称等。代码如下：

```
/*******************************************************************************
 * 名称：thread_init()
 * 功能：互斥量线程初始化
 ******************************************************************************/
int thread_init(void)
{
    rt_err_t result = RT_EOK;
    mutex = rt_mutex_create("mut",                          //互斥量名称
                        RT_IPC_FLAG_FIFO);                  //先进先出模式
    if(mutex == RT_NULL)
        return -1;
    thread1 = rt_thread_create("thread1",                   //线程 1 名称
                        thread1_entry,                      //线程 1 入口函数
                        RT_NULL,                            //线程 1 入口函数参数
                        512,                                //线程 1 内存大小
                        5,                                  //线程 1 优先级
                        20);                                //线程 1 时间片
    if(thread1 == RT_NULL)
        return -1;
    result |= rt_thread_startup(thread1);                   //启动线程 1
    thread2 = rt_thread_create("thread2",                   //线程 2 名称
                        thread2_entry,                      //线程 2 入口函数
                        RT_NULL,                            //线程 2 入口函数参数
                        512,                                //线程 2 内存大小
                        4,                                  //线程 2 优先级
                        20);                                //线程 2 时间片
    if(thread2 == RT_NULL)
        return -1;
    result |= rt_thread_startup(thread2);                   //启动线程 2
    return result;                                          //返回创建及启动结果
}
```

3）两个线程入口函数（zonesion/app/mutex_sample.c）

线程 1：对 num1 和 num2 自加 1，中间加入 100 ms 的挂起延时，如果有互斥量，则线程 1 可以先执行完毕，再执行线程 2。

线程 2：判断 num1 和 num2 是否相等，如果相等则输出互斥量生效信息，否则输出互斥量无效信息。

```
/*******************************************************************************
 * 名称：thread1_entry (void *parameter)
 * 功能：线程 1 入口函数
 * 参数：parameter 表示入口函数参数，在线程创建时传入
 ******************************************************************************/
static void thread1_entry(void *parameter)
```

```
{
    (void)parameter;
    while(1)
    {
        //rt_mutex_take(mutex, RT_WAITING_FOREVER);        //等待获取互斥量
        num1++;                                             //变量自加，用于验证互斥量是否生效
        rt_thread_mdelay(100);                              //延时 100 ms
        num2++;
        //rt_mutex_release(mutex);                          //释放互斥量
    }
}
/*******************************************************************************
* 名称：thread2_entry (void *parameter)
* 功能：线程 2 入口函数
* 参数：parameter 表示入口函数参数，在线程创建时传入
*******************************************************************************/
static void thread2_entry(void *parameter)
{
    (void)parameter;
    while(1)
    {
        rt_mutex_take(mutex, RT_WAITING_FOREVER);          //等待获取互斥量
        if(num1 != num2)                                    //如果两个值相等，则表示互斥量生效
        {
            rt_kprintf("no protect , num1 = %d, num2 = %d\n", num1, num2);
        }
        else
        {
            rt_kprintf("mutex protect ,num1 = num2 = %d\n", num1);
        }
        num1++;
        num2++;
        rt_mutex_release(mutex);                            //释放互斥量
        rt_thread_mdelay(100);                             //延时 100 ms
        if(num1>=50)
            return;
    }
}
```

2.4.3　开发步骤与验证

2.4.3.1　硬件部署

同 1.2.3.1 节。

2.4.3.2　工程调试

（1）将本项目的工程（08-Mutex）文件夹复制到 RT-ThreadStudio\workspace 目录下。

（2）其余同 2.1.3.2 节。

本节主要演示互斥量的作用，需要将 mutex_sample.c 文件（zonesion/app/mutex_sample.c）中 rt_mutex_take() 和 rt_mutex_release() 函数前的注释符删掉。

2.4.3.3　验证效果

（1）关闭 RT-Thread Studio，拔掉仿真器，按下 ZI-ARMEmbed 上的电源按键重新上电。

（2）如果不注释掉线程 1 中 rt_mutex_take()、rt_mutex_release() 函数（允许互斥量的获取和释放），则线程 1 获取到互斥量，线程 1 会将其优先级提升到线程 2 的同等优先级，此时线程 2 不会抢占线程 1，MobaXterm 串口终端 FinSH 控制台输出的信息表明 num1 和 num2 相等。

```
 \ | /
- RT -     Thread Operating System
 / | \     4.1.0 build Oct 14 2022 09:22:26
 2006 - 2022 Copyright by RT-Thread team
msh >mutex protect ,num1 = num2 = 1
mutex protect ,num1 = num2 = 3
mutex protect ,num1 = num2 = 5
mutex protect ,num1 = num2 = 7
mutex protect ,num1 = num2 = 9
mutex protect ,num1 = num2 = 11
mutex protect ,num1 = num2 = 13
mutex protect ,num1 = num2 = 15
mutex protect ,num1 = num2 = 17
mutex protect ,num1 = num2 = 19
mutex protect ,num1 = num2 = 21
mutex protect ,num1 = num2 = 23
mutex protect ,num1 = num2 = 25
mutex protect ,num1 = num2 = 27
mutex protect ,num1 = num2 = 29
mutex protect ,num1 = num2 = 31
mutex protect ,num1 = num2 = 33
mutex protect ,num1 = num2 = 35
mutex protect ,num1 = num2 = 37
mutex protect ,num1 = num2 = 39
mutex protect ,num1 = num2 = 41
mutex protect ,num1 = num2 = 43
mutex protect ,num1 = num2 = 45
mutex protect ,num1 = num2 = 47
mutex protect ,num1 = num2 = 49
```

（3）如果将线程 1 中 rt_mutex_take()、rt_mutex_release() 函数注释掉（禁止互斥量的获取和释放），则线程 1 没有获取到互斥量，由于线程 2 的优先级高于线程 1，因此线程 2 会抢占线程 1，MobaXterm 串口终端 FinSH 控制台输出的信息表明 num1 和 num2 不相等。

```
 \ | /
- RT -     Thread Operating System
```

```
    /|\        4.1.0 build Oct 14 2022 09:22:26
   2006 - 2022 Copyright by RT-Thread team
no protect , num1 = 1, num2 = 0
msh >no protect , num1 = 3, num2 = 2
no protect , num1 = 5, num2 = 4
no protect , num1 = 7, num2 = 6
no protect , num1 = 9, num2 = 8
no protect , num1 = 11, num2 = 10
no protect , num1 = 13, num2 = 12
no protect , num1 = 15, num2 = 14
no protect , num1 = 17, num2 = 16
no protect , num1 = 19, num2 = 18
no protect , num1 = 21, num2 = 20
no protect , num1 = 23, num2 = 22
no protect , num1 = 25, num2 = 24
no protect , num1 = 27, num2 = 26
no protect , num1 = 29, num2 = 28
no protect , num1 = 31, num2 = 30
no protect , num1 = 33, num2 = 32
no protect , num1 = 35, num2 = 34
no protect , num1 = 37, num2 = 36
no protect , num1 = 39, num2 = 38
no protect , num1 = 41, num2 = 40
no protect , num1 = 43, num2 = 42
no protect , num1 = 45, num2 = 44
no protect , num1 = 47, num2 = 46
no protect , num1 = 49, num2 = 48
```

2.4.4　小结

本节主要介绍互斥量的基本概念、工作原理和管理方式。通过本节的学习，读者可掌握
RT-Thread 互斥量的应用。

2.5　RT-Thread 事件集应用开发

在多线程编程中，事件集（Event Set）可以用于线程间的通信和同步。事件集通常包含
一组事件，而线程可以等待其中的一个或多个事件的发生。这种机制有助于线程之间的协同
工作和同步操作。

本节的要求如下：

- ➲ 了解事件集的基本概念。
- ➲ 学习事件集的工作原理。
- ➲ 掌握事件集的管理方式。
- ➲ 掌握 RT-Thread 事件集的应用。

2.5.1   原理分析

2.5.1.1   事件集

事件集也是线程间同步的机制之一，一个事件集可以包含多个事件，利用事件集可以完成一对多、多对多的线程间同步。下面以乘坐公交车为例对事件集进行说明，在公交车站等候公交车时可能有以下几种情况：

（1）P1（第一个乘客）乘坐公交车去某地，只有一种公交车可以到达目的地，等到该公交车即可出发。

（2）P1 乘坐公交车去某地，有三种公交车都可以到达目的地，等到其中任意一辆公交车即可出发。

（3）P1 约 P2（第二个乘客）一起去某地，则 P1 必须要等到"同伴 P2 到达公交车站"与"公交车到达公交车站"两个条件都满足后，才能出发。

这里，可以将 P1 去某地看成线程，将"公交车到达公交车站""同伴 P2 到达公交车站"看成事件的发生，情况（1）是特定事件唤醒线程，情况（2）是任意单个事件唤醒线程，情况（3）是多个事件同时发生才唤醒线程。

2.5.1.2   事件集的工作原理

事件集主要用于线程间的同步，与信号量不同，它的特点是可以实现一对多、多对多的线程间同步。一个线程与多个事件的关系可设置为：其中任意一个事件发生后唤醒线程，或几个事件都发生后才唤醒线程；同样，事件也可以是与多个线程同步的多个事件。这种多个事件的集合可以用一个 32 位无符号整型变量来表示，变量的每一位代表一个事件，线程通过"逻辑与"或"逻辑或"将一个或多个事件关联起来，形成事件组合。事件的"逻辑或"也称为独立型同步，是指线程与任何事件之一同步；事件"逻辑与"称为关联型同步，是指线程与若干事件同步。

在 RT-Thread 中，每个线程都拥有一个事件信息标志（见图 2.22），它有三个属性，分别是 RT_EVENT_FLAG_AND（逻辑与）、RT_EVENT_FLAG_OR（逻辑或）和 RT_EVENT_FLAG_CLEAR（清除标志）。当线程等待事件同步时，可以通过 32 个事件标志和事件信息标志来判断当前接收的事件是否满足同步条件。

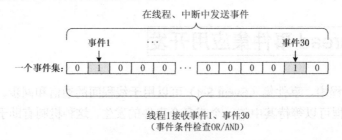

图 2.22   事件信息标志

在图 2.22 中，线程 1 的事件标志中的第 1 位和第 30 位被置 1，如果事件信息标志设为"逻辑与"，则表示线程 1 只有在事件 1 和事件 30 都发生时才会被唤醒；如果事件信息标志设为"逻辑或"，则事件 1 或事件 30 中的任意一个发生时都唤醒线程 1。如果事件信息标志设置了清除标志位，则在线程 1 唤醒后需要主动将事件 1 和事件 30 清 0，否则事件标志将依然存在（即置 1）。

事件集与线程之间的关联如下：

（1）线程等待事件：一个线程可以等待事件集中的一个或多个事件。当线程等待事件时，它会被阻塞直到其中一个事件发生为止。这种机制可用于线程同步，确保线程在条件满足时才继续执行。

（2）事件通知线程：当某个条件满足时，一个线程可以唤醒，或者通知事件集中的某个事件，使得等待该事件的线程可以继续执行。这种方式允许线程之间的协同工作，一个线程的活动可以触发其他线程的响应。

（3）信号量与事件集：在某些情况下，信号量可以被看成一种特殊形式的事件集。信号量通常是一个整数，表示可用资源的数量。当信号量的值大于零时，线程可以执行；当信号量的值为零时，线程可能需要等待。信号量的操作也可以被看成一种事件集的操作。

（4）多线程协作：通过事件集可实现多线程间的协作。例如，在生产者-消费者模型中，一个线程可以等待生产者线程通知有新的数据可用。这样的机制可确保在满足条件时线程协同工作，而不是轮询或等待。

（5）异步编程：在异步编程模型中，事件集也扮演着重要的角色。线程可以注册回调函数，当某个异步操作完成时，相应的事件就会被触发，通知线程继续执行相应的处理。

2.5.1.3　事件集的管理方式

事件集控制块包含了与事件集相关的重要参数，这些参数在事件集功能的实现中具有重要的作用。事件集控制块的相关函数如图 2.23 所示

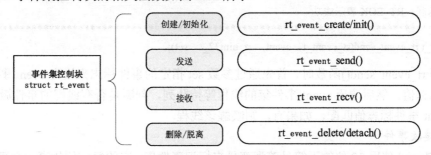

图 2.23　事件集控制块

1）创建事件集

当创建一个事件集时，RT-Thread 会首先创建一个事件集控制块，然后对事件集控制块进行初始化。通过下面的函数可创建事件集：

```
/*******************************************************************************
 * 名称：rt_event_create()
 * 功能：创建事件集
 * 参数：name 表示事件集的名称；flag 表示事件集的标志，其取值为 RT_IPC_FLAG_FIFO 或 RT_
IPC_FLAG_PRIO
 * 返回：RT_NULL 表示创建失败；事件对象集句柄表示创建成功
 *******************************************************************************/
rt_event_t rt_event_create(const char*  name,  rt_uint8_t  flag);
```

在调用 rt_event_create()函数时，RT-Thread 会在对象管理器中分配事件集对象并初始化

这个对象，然后初始化父类 IPC 对象。

2）删除事件集

当不再使用由 rt_event_create()函数创建的事件集时，通过删除事件集可释放系统资源。通过下面的函数可删除事件集：

```
/*****************************************************************************
 * 名称：rt_event_delete()
 * 功能：删除事件集
 * 参数：event 表示事件集对象句柄
 * 返回：RT_EOK 表示成功
 *****************************************************************************/
rt_err_t rt_event_delete(rt_event_t   event);
```

在调用 rt_event_delete()函数删除事件集时，应该确保该事件集不再被使用。在删除事件集前，RT-Thread 会先唤醒所有等待该事件集的线程（线程的返回值是"-RT_ERROR"），然后释放事件集占用的内存。

3）发送事件

通过下面的函数可发送事件集中的一个或多个事件：

```
/*****************************************************************************
 * 名称：rt_event_send()
 * 功能：发送事件
 * 参数：event 表示事件集对象句柄；set 表示发送的一个或多个事件标志
 * 返回：RT_EOK 表示成功
 *****************************************************************************/
rt_err_t rt_event_send(rt_event_t   event,   rt_uint32_t   set);
```

使用 rt_event_send()函数时，首先通过参数 set 指定的事件标志来设定 event 事件集对象的事件标志值，然后遍历 event 事件集的线程等待队列，判断是否有线程的事件激活要求与当前 event 事件标志值匹配，如果有，则唤醒该线程。

4）接收事件

RT-Thread 使用 32 位的无符号整型变量来标识事件集，它的每一位代表一个事件，因此一个事件集可同时等待接收 32 个事件。通过"逻辑与"或"逻辑或"可以选择如何激活线程，使用"逻辑与"表示只有当所有等待的事件都发生时才激活线程，使用"逻辑或"表示只要有一个等待的事件发生就激活线程。使用下面的函数可接收事件：

```
/*****************************************************************************
 * 名称：rt_event_recv()
 * 功能：接收事件
 * 参数：event 表示事件集对象句柄；set 表示接收线程感兴趣的事件；option 表示接收选项；timeout
表示超时时间；recved 表示指向接收到的事件的指针
 * 返回：-RT_EOK 表示成功；-RT_ETIMEOUT 表示超时；-RT_ERROR 表示错误
 *****************************************************************************/
rt_err_t rt_event_recv(rt_event_t   event, rt_uint32_t   set, rt_uint8_t   option, rt_int32_t   timeout,
                rt_uint32_t*   recved);
```

在调用 rt_event_recv()函数时，首先根据 set 参数和接收选项 option 来判断要接收的事件是否发生，如果已经发生则根据参数 option 是否设置 RT_EVENT_FLAG_CLEAR 来决定是否重置相应的事件标志，然后返回（其中 recved 参数返回接收到的事件）；如果没有发生，则把等待的 set 和 option 参数填入线程本身的结构中，然后把线程挂起在此事件上，直到其等待的事件满足条件或等待时间超过指定的超时时间。如果超时时间设置为零，则表示当线程要接收的事件没有满足要求时不等待，直接返回-RT_ETIMEOUT。option 的取值为 RT_EVENT_FLAG_OR 表示选择"逻辑或"的方式接收事件，取值为 RT_EVENT_FLAG_AND 表示选择"逻辑与"的方式接收事件，取值为 RT_EVENT_FLAG_CLEAR 表示清除事件标志。

2.5.2 开发设计与实践

2.5.2.1 硬件设计

本节将 KEY 线程中按键事件发送到事件集，然后在 LED 线程中通过设置感兴趣的事件来决定是否获取事件集，通过事件标志来点亮不同的 LED。KEY、ZI-ARMEmbed 和 LED 的连接如图 2.24 所示。

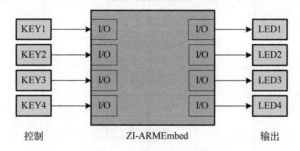

图 2.24 KEY、ZI-ARMEmbed 和 LED 的连接

1）ZI-ARMEmbed 和 LED 的连接
ZI-ARMEmbed 和 LED 的连接请参考 1.5.2.1 节。

2）ZI-ARMEmbed 和 KEY 的连接
ZI-ARMEmbed 和 KEY 的连接如图 2.25 所示。当按键按下时，KEY 的电位是低电位；当按键松开时，KEY 的电位是高电位。本节在按键按下时发送一个按键事件。

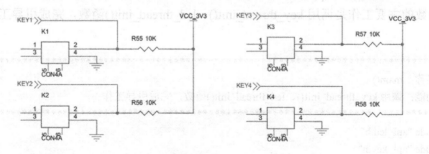

图 2.25 ZI-ARMEmbed 和 KEY 的连接

2.5.2.2 软件设计

软件设计流程如图 2.26 所示。
（1）初始化 LED 线程，初始化 KEY 线程，在 KEY 线程中创建事件集。

（2）编写 KEY 线程的入口函数，在 KEY 线程中将按键事件发送到事件集，在 LED 线程中获取事件集。

（3）按下不同的按键时发送不同的事件，点亮不同的 LED。按下 KEY1 时点亮 LED1，再次按下 KEY1 时熄灭 LED1，其他 KEY 与 LED 的关系与此类似。

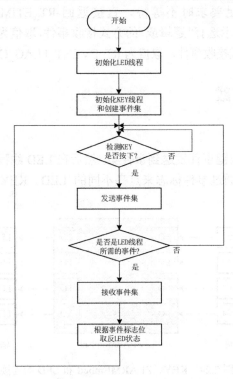

图 2.26   软件设计流程

2.5.2.3　功能设计与核心代码设计

本节需要创建 2 个线程，在 KEY 线程中发送事件，在 LED 线程中接收事件并点亮或熄灭 LED。

1）主函数（zonesion/app/main.c）

主函数的主要工作是调用 key_thread_init()、led_thread_init()函数，完成引导工作。代码如下：

```
/********************************************************************
 * 名称：main()
 * 功能：调用 key_thread_init()、led_thread_init()函数，完成引导工作
 /********************************************************************
#include "apl_led.h"
#include "apl_key.h"
int main(void)
{
    key_thread_init();                              //KEY 线程初始化
    led_thread_init();                              //LED 线程初始化
    return 0;
}
```

2）KEY 线程和事件集初始化函数（zonesion/app/apl_key.c）

KEY 线程和事件集初始化函数的主要工作是在 KEY 线程的初始化中创建事件集。代码如下：

```
/*********************************************************************************
 * 名称：key_thread_init()
 * 功能：KEY 线程初始化
 * 返回：-1 表示事件集创建失败；0 表示事件集创建并线程启动成功；其他值表示创建成功但线程 KEY
 启动失败
 *********************************************************************************/
int key_thread_init(void)
{
    key_event = rt_event_create("key",                      //事件集名称
                        RT_IPC_FLAG_FIFO);                   //选择先进先出模式
    if(key_event == RT_NULL)
        return -1;
    key_thread = rt_thread_create("key",                    //线程名称
                        key_thread_entry,                    //线程入口函数
                        RT_NULL,                             //线程入口函数参数
                        512,                                 //线程内存大小
                        8,                                   //线程优先级
                        20);                                 //线程时间片
    if(key_thread == RT_NULL)
        return -1;
    return rt_thread_startup(key_thread);                   //启动 KEY 线程
}
```

3）KEY 线程入口函数（zonesion/app/apl_key.c）

KEY 线程入口函数的主要工作是完成 KEY 引脚的初始化，如果有按键事件，则 KEY 线程就会将其发送到事件集，事件集的值等于按键的键值。代码如下：

```
/*********************************************************************************
 * 名称：key_thread_entry()
 * 功能：KEY 线程入口函数
 * 参数：*parameter 表示入口函数参数
 *********************************************************************************/
static void key_thread_entry(void *parameter)
{
    (void)parameter;
    unsigned char keyVal = 0;
    key_pin_init();
    while(1)
    {
        if(keyVal != 0)
        {
            if(key_getState() == 0)                         //等待按键弹起
            {
                rt_event_send(key_event, keyVal);          //发送按键事件
```

```
                    keyVal = 0;
                }
            }
            else
                keyVal = key_getState();                    //检测是否有按键按下
            rt_thread_mdelay(100);                          //延时 100 ms
        }
    }
```

4）LED 线程初始化函数（zonesion/app/apl_led.c）

LED 线程初始化函数的主要工作是完成 LED 线程的创建并启动线程。代码如下：

```
/*******************************************************************************
* 名称：led_thread_init()
* 功能：LED 线程初始化函数
* 返回：-1 表示 LED 线程创建失败；0 表示 LED 线程创建并启动成功；其他值表示 LED 线程创建成
功但启动失败
*******************************************************************************/
int led_thread_init(void)
{
    led_thread = rt_thread_create("led",                    //线程名称
                            led_thread_entry,                //线程入口函数
                            RT_NULL,                         //线程入口函数参数
                            256,                             //线程内存大小
                            10,                              //线程优先级
                            20);                             //线程时间片
    if(led_thread == RT_NULL)
        return -1;
    return rt_thread_startup(led_thread);
}
```

5）LED 线程入口函数（zonesion/app/apl_led.c）

LED 线程入口函数的主要工作是完成 LED 引脚的初始化，在主循环中接收事件，根据设置的事件标志来判断是否激活线程，然后输出不同的 LED 状态。代码如下：

```
/*******************************************************************************
* 名称：led_thread_entry()
* 功能：LED 线程入口函数
* 参数：*parameter 表示入口函数参数
*******************************************************************************/
static void led_thread_entry(void *parameter)
{
    (void)parameter;
    rt_err_t result = RT_EOK;
    rt_uint32_t eventVal = 0;
    led_pin_init();                                          //LED 引脚初始化
    while(1)
    {
        result = rt_event_recv(key_event,
            KEY1_NUM | KEY2_NUM | KEY3_NUM | KEY4_NUM,       //接收 4 个按键事件
```

```
                    RT_EVENT_FLAG_OR|RT_EVENT_FLAG_CLEAR,//任意事件发生，完成后清除事件标志
                    RT_WAITING_FOREVER,                     //一直等待
                    &eventVal);                             //事件值
            if(result == RT_EOK)
            {
                if(eventVal & KEY1_NUM) {
                    rt_pin_write(LED1_PIN_NUM, !rt_pin_read(LED1_PIN_NUM));
                }
                if(eventVal & KEY2_NUM) {
                    rt_pin_write(LED2_PIN_NUM, !rt_pin_read(LED2_PIN_NUM));
                }
                if(eventVal & KEY3_NUM) {
                    rt_pin_write(LED3_PIN_NUM, !rt_pin_read(LED3_PIN_NUM));
                }
                if(eventVal & KEY4_NUM) {
                    rt_pin_write(LED4_PIN_NUM, !rt_pin_read(LED4_PIN_NUM));
                }
            }
        }
    }
```

2.5.3　开发步骤与验证

2.5.3.1　硬件部署

同 1.2.3.1 节。

2.5.3.2　工程调试

（1）将本项目的工程（09-Event）文件夹复制到 RT-ThreadStudio\workspace 目录下。

（2）其余同 2.1.3.2 节。

2.5.3.3　验证效果

（1）关闭 RT-Thread Studio，拔掉仿真器，按下 ZI-ARMEmbed 上的电源按键重新上电。

（2）按下 KEY1（KEY1 事件将发送到事件集），点亮 LED1；再次按下 KEY1（KEY1
事件取反后发送到事件集），熄灭 LED1，如图 2.27 和图 2.28 所示。

图 2.27　按下 KEY1 点亮 LED1　　　　　　　图 2.28　再次按下 KEY1 熄灭 LED1

（3）按下 KEY2（KEY2 事件将发送到事件集），点亮 LED2；再次按下 KEY2（KEY2
事件取反后发送到事件集），熄灭 LED2，如图 2.29 和图 2.30 所示。

图 2.29　按下 KEY2 点亮 LED2　　　　图 2.30　再次按下 KEY2 熄灭 LED2

2.5.4　小结

本节主要介绍事件集的基本概念、工作原理和管理方式。通过本节的学习，读者可掌握 RT-Thread 事件集的应用。

2.6 RT-Thread 邮箱应用开发

在多线程编程中，线程之间的通信可以通过各种方式实现，其中一种常见的方式就是使用邮箱（Mailbox）或消息队列（Message Queue）。邮箱充当了线程之间传递消息的容器，其中一个线程可以将消息发送到邮箱，而另一个线程则可以从邮箱中接收并处理消息。邮箱通常用于在线程之间实现异步通信。

本节的要求如下：

- ➲ 了解邮箱的基本概念。
- ➲ 学习邮箱的工作原理。
- ➲ 掌握邮箱的管理方式。
- ➲ 掌握 RT-Thread 邮箱的应用。

2.6.1　原理分析

2.6.1.1　邮箱

邮箱服务是实时操作系统中一种典型的线程间通信方式。例如，有两个线程，线程 1 检测按键状态并发送，线程 2 读取按键状态并根据按键的状态相应地改变 LED 的亮灭。这里就可以使用邮箱进行通信，线程 1 将按键状态作为邮件发送到邮箱，线程 2 在邮箱中读取邮件获得按键状态并对 LED 执行亮灭操作。

这里的线程 1 也可以扩展为多个线程。例如，有 3 个线程，线程 1 检测并发送按键状态，线程 2 检测并发送 ADC 采样信息，线程 3 则根据接收到的信息执行不同的操作。

2.6.1.2　邮箱的工作原理

RT-Thread 邮箱用于线程间通信，开销低、效率高。邮箱的工作原理如图 2.31 所示，线程或中断服务程序把一封 4 B 的邮件发送到邮箱中，而一个或多个线程可以从邮箱中接收这些邮件并进行处理。

图 2.31　邮箱的工作原理

以非阻塞方式发送的邮件能够安全地应用于中断服务程序中，是线程、中断服务程序、定时器向线程发送消息的有效手段。邮件的收取过程可能是阻塞式的，这取决于邮箱中是否有邮件，以及收取邮件时设置的超时时间。当邮箱中不存在邮件且超时时间不为 0 时，邮件收取过程将变成阻塞式的。在这种情况下，只能由线程完成邮件的收取。

当一个线程向邮箱发送邮件时，如果邮箱没满，则把邮件复制到邮箱中。如果邮箱已满，则发送线程可以设置超时时间，选择挂起等待或直接返回"-RT_EFULL"。如果发送线程选择挂起等待，那么当邮箱中的邮件被收取而空出空间来时，挂起等待的发送线程将被唤醒继续发送邮件。

当一个线程从邮箱中接收邮件时，如果邮箱是空的，则接收线程可以选择是否挂起等待直到收到新的邮件而唤醒为止。接收线程也可以设置超时时间，当达到设置的超时时间时，若邮箱依然未收到邮件，则这个选择超时等待的线程将被唤醒并返回"-RT_ETIMEOUT"。如果邮箱中存在邮件，那么接收线程将邮箱中的邮件复制到接收缓冲区中。

使用邮箱进行线程间通信的一般步骤为：

（1）创建邮箱：在程序中创建一个用于存储消息（邮件）的数据结构，如队列，这个数据结构是线程安全的，以确保多个线程可以安全地访问它。

（2）线程发送消息：一个线程在需要与其他线程通信时，可以将邮件封装并发送到邮箱中。这个过程是非阻塞式的，发送线程不必等待接收线程处理邮件。

（3）线程接收消息：另一个线程可以在需要时从邮箱中接收邮件。如果邮箱为空，则线程可能会等待，直到有邮件可用为止。

（4）处理消息：接收线程接收到邮件后，可以解析邮件并执行相应的操作。这个过程可以包含各种线程间协作的逻辑。

2.6.1.3　邮箱的管理方式

邮箱控制块是一个结构体，其中包含了与事件相关的重要参数，在邮箱的功能设计与核心代码设计中起着重要的作用。邮箱控制块的相关函数如图 2.32 所示。

图 2.32　邮箱控制块的相关函数

1）创建邮箱

通过下面的函数可动态地创建一个邮箱：

```
/************************************************************************
 * 名称：rt_mb_create()
 * 功能：创建邮箱
 * 参数：name 表示邮箱名称；size 表示邮箱容量；flag 表示邮箱标志，其取值为 RT_IPC_FLAG_FIFO
或 RT_IPC_FLAG_PRIO
 * 返回：RT_NULL 表示创建失败；邮箱对象句柄表示创建成功
 ************************************************************************/
rt_mailbox_t rt_mb_create (const char*    name, rt_size_t    size, rt_uint8_t    flag);
```

在调用 rt_mb_create()函数创建邮箱时，RT-Thread 会先在对象管理器中分配一个邮箱对象，然后为邮箱对象动态分配一块内存空间用来存放邮件，这块内存的大小等于邮件大小（4 B）与邮箱容量的乘积，最后初始化接收邮件数目和发送邮件在邮箱中的偏移量。

2）删除邮箱

当不再使用由 rt_mb_create()创建的邮箱时，应该删除邮箱以便释放相应的系统资源，一旦完成删除操作，邮箱将被永久性地删除。通过下面的函数可删除邮箱：

```
/************************************************************************
 * 名称：rt_mb_delete()
 * 功能：删除邮箱
 * 参数：mb 表示邮箱对象句柄
 * 返回：RT_EOK 表示成功
 ************************************************************************/
rt_err_t rt_mb_delete (rt_mailbox_t    mb);
```

在删除邮箱时，如果有线程被挂起等待该邮箱，则 RT-Thread 会先唤醒挂起等待该邮箱上的所有线程（线程返回值是"-RT_ERROR"），然后释放邮箱使用的内存，最后删除邮箱。

3）发送邮件

线程或者中断服务程序可以通过邮箱给其他线程发送邮件。通过下面的函数可发送邮件：

```
/************************************************************************
 * 名称：rt_mb_send()
 * 功能：发送邮件
 * 参数：mb 表示邮箱对象句柄；value 表示邮件内容
 * 返回：RT_EOK 表示发送成功；-RT_EFULL 表示邮箱已经满了
 ************************************************************************/
rt_err_t rt_mb_send (rt_mailbox_t    mb,    rt_uint32_t    value);
```

发送的邮件可以是 32 位任意格式的数据，如一个整型变量或者一个指向缓冲区的指针。当邮箱中的邮件已满时，发送邮件的线程或者中断服务程序会收到返回值"-RT_EFULL"。

4）接收邮件

只有当邮箱中有邮件时，接收线程才能立即获取邮件并返回"RT_EOK"，否则接收线程会根据设置的超时时间，加入该邮箱的线程等待队列或直接返回。通过下面的函数可接收邮件：

```
/*************************************************************************
 * 名称：rt_mb_recv()
 * 功能：接收邮件
 * 参数：mb 表示邮箱对象句柄；value 表示邮件内容；timeout 表示超时时间
 * 返回：RT_EOK 表示接收成功；-RT_ETIMEOUT 表示超时；-RT_ERROR 表示失败
 *************************************************************************/
rt_err_t rt_mb_recv (rt_mailbox_t    mb, rt_uint32_t*  value, rt_int32_t    timeout);
```

在接收邮件时，接收线程需要指定接收邮件的邮箱句柄，并指定邮件的存放位置，以及最多能够等待的时间。如果设置了超时时间并且在指定的时间内依然未收到邮件，则返回"-RT_ETIMEOUT"。

2.6.2　开发设计与实践

2.6.2.1　硬件设计

本节的硬件设计同 1.5.2.1 节。本节通过 MobaXterm 串口终端 FinSH 控制台控制发送邮件的内容，LED 线程根据接收到的邮件内容来点亮不同的 LED。

2.6.2.2　软件设计

软件设计流程如图 2.33 所示。

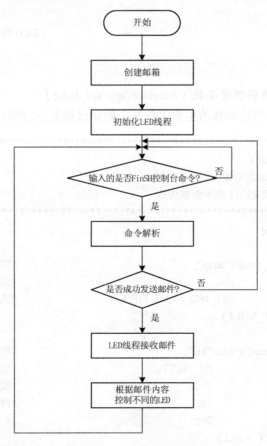

图 2.33　软件设计流程

（1）创建邮箱、初始化 LED 线程。

（2）编写 LED 线程入口函数，在 LED 线程入口函数中完成引脚的初始化，在主循环中不断接收邮箱中的邮件，根据邮件内容点亮不同的 LED。

（3）当在 MobaXterm 串口终端 FinSH 控制台中输入"mailBox send 1"命令时，FinSH 控制台输出"mailBox send success!"，同时点亮 LED1 点亮；当输入"mailBox send 2"命令时，FinSH 控制台输出"mailBox send success!"，同时点亮 LED2；当输入"mailBox send 4"命令时，FinSH 控制台输出"mailBox send success!"，同时点亮 LED3；当输入"mailBox send 8"命令时，FinSH 控制台输出"mailBox send success!"，同时点亮 LED4。

2.6.2.3 功能设计与核心代码设计

通过原理学习控制，一封邮件的大小是 32 bit，发送的邮件可以是 32 bit 的任意格式数据。

1）主函数（zonesion/app/main.c）

主函数的主要工作是调用 led_thread_init()函数，完成引导工作。代码如下：

```
/*******************************************************************************
 * 名称：main()
 * 功能：调用 led_thread_init()函数，完成引导工作
/*******************************************************************************
#include "apl_led.h"
int main(void)
{
    led_thread_init();                                          //LED 线程初始化
    return 0;
}
```

2）线程初始化和邮箱创建函数（zonesion/app/apl_led.c）

线程初始化和邮箱创建函数的主要工作是初始化线程并创建邮箱。代码如下：

```
/*******************************************************************************
 * 名称：led_thread_init()
 * 功能：LED 线程和邮箱初始化
 * 返回：0 表示创建成功；-1 表示创建失败
 *******************************************************************************/
int led_thread_init(void)
{
    temp_mb = rt_mb_create("temp",                              //邮箱名称
                           10,                                  //邮箱大小 10×4 B= 40 B
                           RT_IPC_FLAG_FIFO);                   //选择先进先出模式
    if(temp_mb == RT_NULL)
        return -1;
    led_thread = rt_thread_create("led",                        //线程入口函数
                           RT_NULL,                             //线程入口函数参数
                           256,                                 //线程内存大小
                           10,                                  //线程优先级
                           20);                                 //线程时间片
    if(led_thread == RT_NULL)
        return -1;
```

```
        return rt_thread_startup(led_thread);                         //启动线程
}
```

3）LED 线程入口函数（zonesion/app/apl_led.c）

LED 线程入口函数的主要工作是完成邮件的接收，并根据邮件内容点亮不同的 LED。代码如下：

```
/*******************************************************************************
* 名称: led_thread_entry(void *parameter)
* 功能: LED 线程入口函数
* 参数: parameter 表示入口函数参数，在线程创建时传入
*******************************************************************************/
static void led_thread_entry(void *parameter)
{
    (void)parameter;
    rt_err_t result = RT_EOK;
    char *mbData = RT_NULL;
    led_pin_init();                                         //LED 引脚初始化
    while(1)
    {
        result = rt_mb_recv(temp_mb,                        //邮箱对象句柄
                        (rt_ubase_t*)&mbData,               //邮件内容
                        RT_WAITING_FOREVER);                //一直等待
        if(result == RT_EOK)
        {
            led_ctrl(atoi(mbData));                         //根据邮件内容点亮不同的 LED
        }
    }
}
```

4）自定义 FinSH 控制台命令函数（zonesion/app/finsh_cmd/finsh_ex.c）

自定义 FinSH 控制台命令函数的主要工作是实现邮件发送的命令。代码如下：

```
/*******************************************************************************
* 名称: mailBox (int argc, char **argv)
* 功能: 自定义 FinSH 控制台命令
* 参数: argc 表示参数个数; argv 表示指向参数字符串指针的数组指针
*******************************************************************************/
#include <rtthread.h>
#include "apl_led.h"
int mailBox(int argc, char **argv)
{
    rt_err_t result = RT_EOK;
    if(!rt_strcmp(argv[1], "send"))
    {
        if(argc < 3)
        {
            rt_kprintf("Please input mail content!\n");
```

```
            return 0;
        }
        result = rt_mb_send(temp_mb, (rt_ubase_t)argv[2]);              //发送邮件
        if(result == RT_EOK)
            rt_kprintf("mailBox send success!\n");
        else
            rt_kprintf("mailBox failed to send, errCode:%d\n", result);
    }
    else
        rt_kprintf("Please input 'mailBox <send>'\n");
    return 0;
}
MSH_CMD_EXPORT(mailBox, mailBox sample);
```

2.6.3 开发步骤与验证

2.6.3.1 硬件部署

同 1.2.3.1 节。

2.6.3.2 工程调试

（1）将本项目的工程（10-MailBox）文件夹复制到 RT-ThreadStudio\workspace 目录下。

（2）其余同 2.1.3.2 节。

2.6.3.3 验证效果

（1）关闭 RT-Thread Studio，拔掉仿真器，按下 ZI-ARMEmbed 上的电源按键重新上电。

（2）当 MobaXterm 串口终端 FinSH 控制台中输入"mailBox send 1"命令时，解析自定义 FinSH 控制台命令后，FinSH 控制台发送邮件，LED1 线程获取邮件内容后点亮 LED1，如图 2.34 所示，同时在 FinSH 控制台输出"mailBox send success!"。

```
 \ | /
- RT -     Thread Operating System
 / | \      4.1.0 build Oct 14 2022 11:30:04
 2006 - 2022 Copyright by RT-Thread team
msh >mailBox send 1
mailBox send success!
msh >
```

（3）当 MobaXterm 串口终端 FinSH 控制台中输入"mailBox send 2"命令时，解析自定义 FinSH 控制台命令后，FinSH 控制台发送邮件，LED2 线程获取邮件内容后点亮 LED2，如图 2.35 所示，同时在 FinSH 控制台输出"mailBox send success!"。

其他命令与此类似。

```
 \ | /
- RT -     Thread Operating System
 / | \      4.1.0 build Oct 14 2022 11:30:04
 2006 - 2022 Copyright by RT-Thread team
msh >mailBox send 1
```

```
mailBox send success!
msh >mailBox send 2
mailBox send success!
msh >
```

图 2.34　点亮 LED1　　　　　　　　　　　　　　　　图 2.35　点亮 LED2

2.6.4　小结

本节主要介绍邮箱的基本概念、工作原理和管理方式。通过本节的学习，读者可掌握 RT-Thread 邮箱的应用。

2.7　RT-Thread 消息队列应用开发

在多线程编程中，消息队列是一种常见的、用于线程间通信的机制。线程中的消息队列通常是一个数据结构，用于存储和传递消息，以实现线程之间的解耦和异步通信。

本节的要求如下：

- ☞ 了解消息队列的基本概念。
- ☞ 学习消息队列的工作原理。
- ☞ 掌握消息队列的管理方式。
- ☞ 掌握 RT-Thread 消息队列的应用。

2.7.1　原理分析

2.7.1.1　消息队列

消息队列是一种常用的线程间通信机制，是邮箱的扩展，可以应用在多种场合，如线程间的消息交换、使用串口接收不定长数据等。

线程中消息队列的一般工作流程为：

（1）创建消息队列：在程序中创建一个消息队列数据结构，用于存储消息。这个队列通常是线程安全的，以确保多个线程可以安全地访问消息队列。

（2）发送消息：一个线程在需要与其他线程通信时，可以创建一个消息并将其放入消息队列中。这个过程是非阻塞式的，发送线程不需要等待接收线程处理消息。

（3）接收消息：另一个线程可以在需要时从消息队列中获取消息。如果消息队列为空，则接收线程可能会等待，直到有消息可用为止。

（4）处理消息：接收线程获取到消息后解析消息并执行相应的操作，这个过程可以包含各种线程间协作的逻辑。

2.7.1.2 消息队列的工作原理

消息队列能够接收来自线程或中断服务程序中不固定长度的消息，并把消息缓存在自己的内存空间中。其他线程能够从消息队列中读取相应的消息，当消息队列为空时，可以挂起线程。当有新的消息到达时，挂起的线程将被唤醒以接收并处理消息。消息队列是一种异步的通信方式。

消息队列的工作原理如图 2.36 所示，线程或中断服务程序可以将一条或多条消息放入消息队列中，线程也可以从消息队列中获得消息。当有多个消息发送到消息队列时，通常将先进入消息队列的消息先传给线程，线程先得到的是最先进入消息队列的消息，即遵循先进先出（FIFO）的原则。

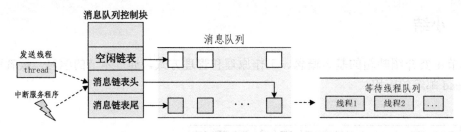

图 2.36　消息队列的工作原理

RT-Thread 的消息队列由多个元素组成。系统在创建消息队列时，为消息队列分配了消息队列控制块，其中包括消息队列名称、内存缓冲区、消息大小和消息队列长度等。每个消息队列中都包含多个消息框，每个消息框都可以存放一条消息。消息队列中的第一个和最后一个消息框分别称为消息链表头和消息链表尾，对应于消息队列控制块中的 msg_queue_head 和 msg_queue_tail。有些消息框可能是空的，它们通过 msg_queue_free 形成一个空闲消息链表。消息队列中消息框的总数是消息队列的长度，这个长度可在创建消息队列时指定。

2.7.1.3 消息队列的管理方式

消息队列控制块是一个结构体，其中包含了与消息队列相关的重要参数，这些参数在消息队列的功能设计与核心代码设计中起着重要的作用。消息队列控制块的相关函数如图 2.37 所示。

图 2.37　消息队列控制块

1）创建消息队列

消息队列在使用前，应该被创建出来，或对已有的静态消息队列对象进行初始化。通过下面的函数可创建消息队列：

```
/************************************************************************
 * 名称：rt_mq_create()
 * 功能：创建消息队列
 * 参数：name 表示消息队列的名称；msg_size 表示消息队列中一条消息的最大长度，单位为字节；
max_msgs 表示消息队列中消息的最大个数；flag 表示消息队列采用的等待方式，其取值为
RT_IPC_FLAG_FIFO 或 RT_IPC_FLAG_PRIO
 * 返回：RT_EOK 表示发送成功；消息队列对象句柄表示成功；RT_NULL 表示失败
 ************************************************************************/
rt_mq_t rt_mq_create(const char* name, rt_size_t msg_size, rt_size_t max_msgs, rt_uint8_t flag);
```

在调用 rt_mq_create() 函数创建消息队列时，RT-Thread 会先在对象管理器中分配一个消息队列对象；然后为消息队列对象分配一块内存空间，组成空闲消息链表，这块内存的大小为 [消息大小 + 消息头（用于链接消息链表）的大小] ×消息队列中消息的最大个数；最后初始化消息队列，此时消息队列是空的。

2）删除消息队列

当不再使用消息队列时，应该删除它以释放系统资源，一旦完成删除操作，消息队列将被永久性地删除。通过下面的函数可删除消息队列：

```
/************************************************************************
 * 名称：rt_mq_delete()
 * 功能：删除消息队列
 * 参数：mq 表示消息队列对象句柄
 * 返回：RT_EOK 表示成功
 ************************************************************************/
rt_err_t rt_mq_delete(rt_mq_t  mq);
```

在调用 rt_mq_delete() 函数删除消息队列时，如果有线程在该消息队列的线程等待队列中，则 RT-Thread 先唤醒该消息队列的线程等待队列中的所有线程（线程返回值是"-RT_ERROR"），然后释放消息队列占用的内存，最后删除消息队列对象。

3）发送消息

线程或者中断服务程序可以给消息队列发送消息。在发送消息时，消息队列先从空闲消息链表上取下一个空闲消息框，把线程或者中断服务程序发送的消息复制到空闲消息框上，然后把该消息框挂到消息队列的尾部。当且仅当空闲消息链表上有可用的空闲消息框时，线程或中断服务程序才能成功发送消息；当空闲消息链表上无可用的空闲消息框时，说明消息队列已满，此时发送消息的线程或者中断服务程序会收到一个错误码（-RT_EFULL）。通过下面的函数可发送消息：

```
/************************************************************************
 * 名称：rt_mq_send()
 * 功能：发送消息
 * 参数：mq 表示消息队列对象句柄；buffer 表示消息内容；size 表示消息大小
 * 返回：RT_EOK 表示成功；-RT_EFULL 表示消息队列已满；-RT_ERROR 表示失败，发送的消息
长度大于消息队列中消息的最大长度
 ************************************************************************/
rt_err_t rt_mq_send (rt_mq_t mq, void* buffer, rt_size_t size);
```

　　在调用 rt_mq_send()函数发送消息时，线程或中断服务程序需要指定消息队列对象句柄（即指向消息队列控制块的指针），并且指定发送的消息内容以及消息大小。在发送一个消息之后，空闲消息链表在消息链表头的消息将被转移到消息链表尾。

　　4）发送紧急消息

　　发送紧急消息的过程与发送普通消息几乎一样，唯一的不同是，当发送紧急消息时，从空闲消息链表上取下来的消息框不是挂到消息链表尾，而是挂到消息链表头，这样线程就能够优先接收到紧急消息，从而及时进行消息处理。通过下面的函数可发送紧急消息：

```
/************************************************************************
 * 名称：rt_mq_urgent()
 * 功能：发送紧急消息
 * 参数：mq 表示消息队列对象句柄；buffer 表示消息内容；size 表示消息大小
 * 返回：RT_EOK 表示成功；-RT_EFULL 表示消息队列已满；-RT_ERROR 表示失败
 ************************************************************************/
rt_err_t rt_mq_urgent(rt_mq_t mq, void* buffer, rt_size_t size);
```

　　在调用 rt_mq_urgent()函数发送紧急消息时，线程或中断服务程序需要指定消息队列对象句柄（即指向消息队列控制块的指针），并且指定发送的紧急消息内容以及紧急消息大小。

　　5）接收消息

　　当消息队列中有消息时，线程才能接收消息，否则线程会根据设定的超时时间，或挂到消息队列的线程等待队列上，或直接返回。通过下面的函数可接收消息：

```
/************************************************************************
 * 名称：rt_mq_recv()
 * 功能：接收消息
 * 参数：mq 表示消息队列对象句柄；buffer 表示消息内容；size 表示消息大小；timeout 表示设置的
超时时间
 * 返回：RT_EOK 表示成功收到；-RT_ETIMEOUT 表示超时；-RT_ERROR 表示失败
 ************************************************************************/
rt_err_t rt_mq_recv (rt_mq_t mq, void* buffer, rt_size_t size, rt_int32_t timeout);
```

2.7.2　开发设计与实践

2.7.2.1　软件设计

软件设计流程如图 2.38 所示。

（1）创建消息队列，在主函数中初始化线程。

（2）编写线程 1 和 2 的入口函数，线程 1 每 500 ms 发送一次消息，线程 2 每 750 ms 接收一次消息。

（3）在 MobaXterm 串口终端 FinSH 控制台中输入命令，通过命令来发送紧急消息。当输入"msgq urgent 9999"时，线程 2（接收线程）立即输出"thread2: recv message number: 9999"。

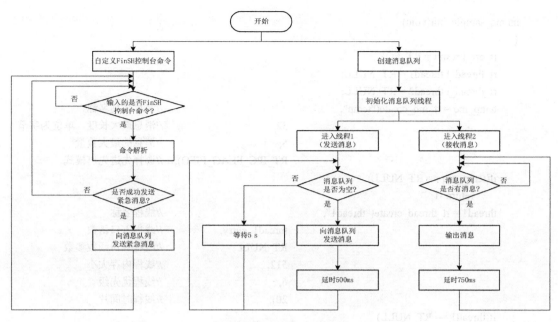

图 2.38　软件设计流程

2.7.2.2　功能设计与核心代码设计

通过原理和软件设计流程的学习，代码的设计就变得很清晰了。首先在 main 函数中调用消息队列线程初始化函数，然后在消息队列线程初始化函数中创建 2 个线程。另外，还需要自定义 MobaXterm 串口终端 FinSH 控制台命令，方便发送紧急消息。在 2 个线程中，一个用于发送消息，另一个用于接收消息。

1）主函数（zonesion/app/main.c）

主函数的主要工作是调用 mq_sample_init()函数，完成引导工作。代码如下：

```
/*******************************************************************************
* 名称：main()
* 功能：调用 mq_sample_init()函数，完成引导工作
* 返回：0 表示成功
*******************************************************************************/
#include "mq_sample.h"
int main(void)
{
    mq_sample_init();                                    //消息队列线程初始化
    return 0;
}
```

2）消息队列线程初始化函数（zonesion/app/mq_sample.c）

消息队列线程初始化函数的主要工作是初始化消息队列，并创建 2 个线程。代码如下：

```
/*******************************************************************************
* 名称：mq_sample_init()
* 功能：消息队列线程初始化
* 返回：0 表示创建成功；-1 表示创建失败
*******************************************************************************/
```

```
int mq_sample_init(void)
{
    rt_err_t result = RT_EOK;
    rt_thread_t thread1 = RT_NULL;
    rt_thread_t thread2 = RT_NULL;
    temp_mq = rt_mq_create("temp",                    //消息队列名称
                           32,                        //消息最大长度，单位为字节
                           5,                         //消息的最大数量
                           RT_IPC_FLAG_FIFO);         //选择先进先出模式
    if(temp_mq == RT_NULL)
        return -1;
    thread1 = rt_thread_create("thread1",             //线程名称
                               thread1_entry,         //线程入口函数
                               RT_NULL,               //线程入口函数参数
                               512,                   //线程内存大小
                               6,                     //线程优先级
                               20);                   //线程时间片
    if(thread1 == RT_NULL)
        return -1;
    result = rt_thread_startup(thread1);              //启动线程
    thread2 = rt_thread_create("thread2",             //线程名称
                               thread2_entry,         //线程入口函数
                               RT_NULL,               //线程入口函数参数
                               512,                   //线程内存大小
                               6,                     //线程优先级
                               20);                   //线程时间片
    if(thread2 == RT_NULL)
        return -1;
    result = rt_thread_startup(thread2);              //启动线程
    return result;                                    //返回创建及启动结果
}
```

3）线程入口函数（zonesion/app/mq_sample.c）

线程 1 的入口函数的主要工作是每 500 ms 向消息队列发送一次消息，当消息队列已满时，等待 5 s。线程 2 的入口函数的主要工作是每 750 ms 从消息队列接收一次消息。代码如下：

```
/********************************************************************************
 * 名称：thread1_entry (void *parameter)
 * 功能：线程 1 入口函数，发送消息
 * 参数：parameter 表示入口函数参数，在线程创建时传入
 ********************************************************************************/
static void thread1_entry(void *parameter)
{
    (void)parameter;
    rt_err_t result = RT_EOK;
    unsigned long num = 0;
    char tempBuf[32] = {0};
    while(1)
```

```
    {
        num++;
        sprintf(tempBuf, "%u", num);                                    //向消息队列发送消息
        result = rt_mq_send(temp_mq,                                     //消息队列对象句柄
                                    tempBuf,                             //消息
                                    sizeof(tempBuf));                    //消息大小
        if(result == RT_EOK)
            rt_kprintf("thread1: send message number: %s\n", tempBuf);
        else if(result == -RT_EFULL)                                    //消息队列已满，等待 5 s
        {
            rt_kprintf("thread1: message queue is full, delay 5 sec!\n");
            rt_thread_mdelay(5000);                                     //延时 5000 ms
        }
        rt_thread_mdelay(500);                                          //延时 500 ms
    }
}
/*****************************************************************************
* 名称：thread2_entry (void *parameter)
* 功能：线程 2 入口函数，接收消息
* 参数：parameter 表示入口函数参数，在线程创建时传入
*****************************************************************************/
static void thread2_entry(void *parameter)
{
    (void)parameter;
    rt_err_t result = RT_EOK;
    char tempBuf[32] ={0};
    while(1)
    {
        result = rt_mq_recv(temp_mq,                                    //消息队列对象句柄
                                    tempBuf,                            //消息
                                    sizeof(tempBuf),                    //消息大小
                                    RT_WAITING_FOREVER);                //一直等待
        if(result == RT_EOK)
        {
            rt_kprintf("thread2: recv message number: %s\n", tempBuf);  //显示接收到的消息序号
        }
        rt_thread_mdelay(750);                                          //延时 750 ms
    }
}
```

4）自定义 FinSH 控制台命令函数（zonesion/app/finsh_cmd/finsh_ex.c）

自定义 FinSH 控制台命令函数的主要工作是输入指定命令后发送紧急消息，消息内容是输入的命令。代码如下：

```
/*****************************************************************************
* 名称：msgq (int argc, char **argv)
* 功能：自定义 FinSH 控制台命令
* 参数：argc 表示参数个数；argv 表示指向参数字符串指针的数组指针
```

```
**************************************************************************/
#include <rtthread.h>
#include "mq_sample.h"
int msgq(int argc, char **argv)
{
    rt_err_t result = RT_EOK;
    if(!rt_strcmp(argv[1], "urgent"))
    {
        if(argc < 3)
        {
            rt_kprintf("Please input message!\n");
            return 0;
        }
        result = rt_mq_urgent(temp_mq, argv[2], sizeof(argv[2]));          //发送紧急消息
        if(result == RT_EOK)
            rt_kprintf("urgent message send success!\n");
        else
            rt_kprintf("urgent message fail to send, errCode:%d\n", result);
    }else
        rt_kprintf("Please input 'msgq <urgent>'\n");
    return 0;
}
MSH_CMD_EXPORT(msgq, message queue sample);
```

2.7.3　开发步骤与验证

2.7.3.1　硬件部署

同 1.2.3.1 节。

2.7.3.2　工程调试

（1）将本项目的工程（11-MessageQueue）文件夹复制到 RT-ThreadStudio\workspace 目录下。

（2）其余同 2.1.3.2 节。

2.7.3.3　验证效果

（1）关闭 RT-Thread Studio，拔掉仿真器，按下 ZI-ARMEmbed 上的电源按键重新上电。

（2）在 MobaXterm 串口终端 FinSH 控制台输出"thread1: send message number: 1   thread2: recv message number: 1"并依次递增数字。线程 1（发送线程）比线程 2（接收线程）的延时更短，所以消息会在消息队列中累积。

```
 \ | /
- RT -     Thread Operating System
 / | \     4.1.0 build Oct 14 2022 11:45:30
 2006 - 2022 Copyright by RT-Thread team
thread1: send message number: 1
thread2: recv message number: 1
```

```
msh >thread1: send message number: 2
thread2: recv message number: 2
thread1: send message number: 3
thread2: recv message number: 3
thread1: send message number: 4
thread1: send message number: 5
thread2: recv message number: 4
thread1: send message number: 6
thread2: recv message number: 5
thread1: send message number: 7
```

（3）线程 1 在消息队列已满时延时 5 s，FinSH 控制台输出"message queue is full, delay 5 sec!"，线程 1（发送线程）停止发送消息，线程 2（接收线程）不断接收消息。

```
thread1: send message number: 33
thread1: message queue is full, delay 5 sec!
thread2: recv message number: 29
thread2: recv message number: 30
thread2: recv message number: 31
thread2: recv message number: 32
thread2: recv message number: 33
thread1: send message number: 35
thread2: recv message number: 35
```

（4）在 MobaXterm 串口终端 FinSH 控制台中输入"msgq urgent 9999"可发送紧急消息，线程 2（接收线程）立即显示"thread2: recv message number: 9999"。

```
urgent message send success!
msh >msgq urgent 9999
thread2: recv message number: 9999
thread1: send message number: 119
thread2: recv message number: 119
thread1: send message number: 120
thread2: recv message number: 120
thread1: send message number: 121
```

2.7.4　小结

本节主要介绍消息队列的基本概念、工作原理和管理方式。通过本节的学习，读者可掌握 RT-Thread 消息队列的应用。

2.8 RT-Thread 信号应用开发

在多线程编程中，线程可以通过信号（Signals）来进行通信。信号是一种轻量级的通知机制，用于通知线程或进程发生了某个特定事件。

本节的要求如下：

- ➲ 了解信号的基本概念。
- ➲ 学习信号的工作原理。
- ➲ 掌握信号的管理方式。
- ➲ 掌握 RT-Thread 信号的应用。

2.8.1 原理分析

2.8.1.1 信号

信号（又称为软中断信号）是在软件层次上对中断机制的一种模拟。从原理上讲，一个线程收到一个信号与处理器收到一个中断请求是类似的。

2.8.1.2 信号的工作原理

信号在 RT-Thread 中用于异步通信，POSIX 标准定义了 sigset_t 类型，该类型可定义一个信号集。sigset_t 类型在不同的系统可能有不同的定义方式。RT-Thread 将 sigset_t 定义成 unsigned long 类型并命名为 rt_sigset_t，应用程序能够使用的信号为 SIGUSR1（10）和 SIGUSR2（12）。

信号本质是软中断，用来通知线程发生了异步事件，可用于线程之间的异常通知、应急处理。一个线程不必通过任何操作来等待信号的到达，事实上，线程也不知道信号到底什么时候到达，线程之间可以互相通过调用 rt_thread_kill()发送软中断信号。

收到信号的线程对信号的处理方法可以分为 3 类：

（1）类似于中断服务程序，对于需要处理的信号，线程可以指定处理函数，由该函数来处理信号。

（2）忽略某个信号，对该信号不做任何处理，就像未发生过一样。

（3）对该信号的处理方式是保留系统的默认值。

信号的工作原理如图 2.39 所示，假设线程 1 需要对信号进行处理，首先线程 1 会安装一个信号并解除阻塞，并在安装信号的同时设定对信号的异常处理方式；然后其他线程可以给线程 1 发送信号，触发线程 1 对该信号的处理。当信号被传递给线程 1 时，如果线程 1 正处于挂起状态，则将线程 1 的状态改为就绪状态，从而去处理对应的信号。如果线程 1 处于运行状态，则会在线程 1 的栈基础上建立新的栈帧空间去处理对应的信号。需要注意的是使用的线程栈大小也会相应增加。

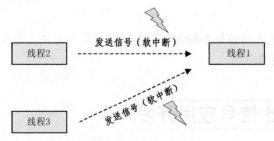

图 2.39 信号的工作原理

2.8.1.3 信号的管理方式

信号控制块的相关函数如图 2.40 所示，对信号的操作主要包括：安装信号、阻塞/解除阻塞信号、发送信号、等待信号。

图 2.40　信号控制块的相关函数

1）安装信号

如果线程要处理某一信号，就要在线程中安装该信号。安装信号的目的是确定信号值和线程针对该信号的动作之间的映射关系，即线程将要处理哪个信号，当该信号被传递给线程时，线程将执行何种操作。通过下面的函数可安装信号：

```
/******************************************************************************
 * 名称：rt_signal_install()
 * 功能：安装信号
 * 参数：signo 表示信号值（只有 SIGUSR1 和 SIGUSR2 是开放给用户使用的）；handler 表示设置对
信号的处理方式
 * 返回：SIG_ERR 表示错误的信号；安装信号前的 handler 表示成功安装信号
 ******************************************************************************/
rt_sighandler_t rt_signal_install(int  signo,  rt_sighandler_t  handler);
```

其中，rt_sighandler_t 是定义信号处理函数的函数指针类型。

在调用 rt_signal_install()函数安装信号时设定的 handler 参数，决定了该信号的不同的处理方法。处理方法可以分为三种：

（1）类似于中断的处理方式，handle 参数指向信号发生时用户自定义的处理函数，由该函数来处理信号。

（2）将 handle 参数设为 SIG_IGN，忽略某个信号，对该信号不做任何处理，就像该信号未发生过一样。

（3）将 handle 参数设为 SIG_DFL，系统会调用默认的处理函数_signal_default_handler()来处理信号。

2）阻塞信号

阻塞信号也可以理解为屏蔽信号。如果某个信号被阻塞，则该信号将不会传递到安装该信号的线程，也不会引发软中断处理。调用下面的函数可阻塞信号：

```
/******************************************************************************
 * 名称：rt_signal_mask()
 * 功能：阻塞信号
 * 参数：signo 表示信号值
 ******************************************************************************/
void rt_signal_mask(int  signo);
```

3）解除阻塞信号

当线程安装了多个信号时，使用下面的函数可以使线程对其中某些信号给予"关注"。

在发送这些被"关注"的信号时会引发该线程的软中断。通过下面的函数可解除阻塞信号：

```
/*******************************************************************************
 * 名称：rt_signal_unmask()
 * 功能：解除信号阻塞
 * 参数：signo 表示信号值
 *******************************************************************************/
void rt_signal_unmask(int   signo);
```

4）发送信号

当需要进行异常处理时，可以给设定了处理异常的线程发送信号。通过下面的函数可发送信号：

```
/*******************************************************************************
 * 名称：rt_thread_kill()
 * 功能：发送信号
 * 参数：tid 表示接收信号的线程；sig 表示信号值
 * 返回：RT_EOK 表示发送成功；-RT_EINVAL 表示参数错误
 *******************************************************************************/
int rt_thread_kill(rt_thread_t   tid,   int   sig);
```

5）等待信号

如果没有等到指定的信号，则将线程会被挂起，直到等到这个信号或者等待时间超过指定的超时时间为止。如果等到了指定的信号，则将指向该信号的指针存入指向信号信息的指针。通过下面的函数可等待信号：

```
/*******************************************************************************
 * 名称：rt_signal_wait()
 * 功能：等待信号
 * 参数：set 表示指定等待的信号；si 表示指向存储等待信号信息的指针；timeout 表示指定的等待时
间，即超时时间
 * 返回：RT_EOK 表示等待信号成功；-RT_ETIMEOUT 表示超时；-RT_EINVAL 表示参数错误
 *******************************************************************************/
int rt_signal_wait(const rt_sigset_t   *set, rt_siginfo_t   *si, rt_int32_t   timeout);
```

其中，rt_siginfo_t 是定义信号信息的数据类型

2.8.2　开发设计与实践

2.8.2.1　硬件设计

本节的硬件设计同 1.5.2.1 节。为了通过发送信号触发操作，本节使用信号触发软中断来控制 LED。

2.8.2.2　软件设计

软件设计流程如图 2.41 所示。

（1）首先创建一个线程，并在 main 函数中完成线程的初始化工作。

（2）编写线程入口函数，在线程中完成信号的安装并解除阻塞。

（3）编写信号的中断处理函数，在处理函数中完成对 LED 的控制，并显示信号值。

（4）在 MobaXterm 串口终端 FinSH 控制台中输入"sig kill"命令，FinSH 控制台输出"signal send success!"，同时依次点亮 4 个 LED，即实现 LED 流水灯效果。

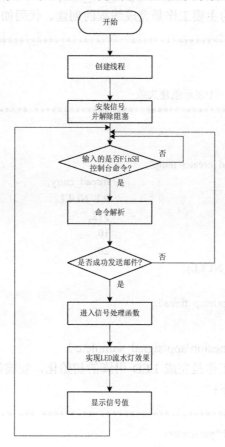

图 2.41　软件设计流程

2.8.2.3　功能设计与核心代码设计

通过原理和软件设计流程的学习，代码的设计就变得很清晰了。首先在主函数中调用 thread_init()函数，并完成引导工作，在信号线程初始化中创建一个线程；然后在该线程中安装信号并解除阻塞；最后在 MobaXterm 串口终端 FinSH 控制台中输入"sig kill"命令，FinSH 控制台输出"signal send success!"，同时依次点亮 4 个 LED。

1）主函数（zonesion/app/main.c）

主函数的主要工作是调用 thread_init()函数，完成引导工作。代码如下：

```
/********************************************************************************
* 名称：main()
* 功能：调用 thread_init()函数，完成引导工作
********************************************************************************/
#include "signal_sample.h"
int main(void)
{
    thread_init();                                          //信号线程初始化
```

```
    return 0;
}
```

2）信号线程初始化函数（zonesion/app/signal_sample.c）

信号线程初始化函数的主要工作是完成线程的创建。代码如下：

```
/*****************************************************************************
* 名称: thread_init()
* 功能: 信号线程初始化
* 返回: 0 表示创建成功, -1 表示创建失败
*****************************************************************************/
int thread_init(void)
{
    temp_thread = rt_thread_create("temp",                  //线程名称
                                   thread_entry,            //线程入口函数
                                   RT_NULL,                 //线程入口函数参数
                                   512,                     //线程内存大小
                                   10,                      //线程优先级
                                   20);                     //线程时间片
    if(temp_thread == RT_NULL)
        return -1;
    return rt_thread_startup(temp_thread);                  //启动线程
}
```

3）线程入口函数（zonesion/app/signal_sample.c）

线程入口函数的主要工作是完成 LED 引脚的初始化，安装信号并解除信号阻塞，之后进入主循环挂起。代码如下：

```
/*****************************************************************************
* 名称: thread_entry(void *parameter)
* 功能: 线程入口函数
* 参数: parameter 表示入口函数参数，在线程创建时传入
*****************************************************************************/
static void thread_entry(void *parameter)
{
    (void)parameter;
    led_pin_init();                                         //LED 引脚初始化
    rt_signal_install(SIGUSR1, signal_handle);             //安装信号
    rt_signal_unmask(SIGUSR1);                             //解除阻塞
    while(1)
    {
        rt_thread_mdelay(1000);                            //延时 1000 ms
    }
}
```

4）信号处理函数（zonesion/app/signal_sample.c）

信号处理函数的主要工作是实现 LED 流水灯效果，并显示信号值。代码如下：

```
/*****************************************************************************
* 名称: signal_handle(int sig)
```

```
* 功能：信号处理函数
* 参数：sig 信号值
*********************************************************************************/
static void signal_handle(int sig)
{
    for(unsigned char i=0x01; i<=0x10; i<<=1)                    //LED 流水灯效果
    {
        led_ctrl(i);                                            //控制 LED
        rt_thread_mdelay(500);                                  //延时 500 ms
    }
    rt_kprintf("thread received signal %d\n", sig);             //显示信号值
}
```

5）自定义 FinSH 控制台命令函数（zonesion/app/finsh_cmd/finsh_ex.c）

自定义 FinSH 控制台命令函数的主要工作是发送信号。代码如下：

```
/*********************************************************************************
* 名称：sig (int argc, char **argv)
* 功能：自定义 FinSH 控制台命令
* 参数：argc 表示参数个数；argv 表示指向参数字符串指针的数组指针
*********************************************************************************/
#include <rtthread.h>
#include "signal_sample.h"

int sig(int argc, char **argv)
{
    rt_err_t result = RT_EOK;
    if(!rt_strcmp(argv[1], "kill"))
    {
        result = rt_thread_kill(temp_thread, SIGUSR1);          //发送信号
        if(result == RT_EOK)
            rt_kprintf("signal send success!\n");
        else
            rt_kprintf("signal failed to send, errCode:%d\n", result);
    }
    else
        rt_kprintf("Please input 'sig <kill>'\n");
    return 0;
}
MSH_CMD_EXPORT(sig, signal sample);
```

2.8.3　开发步骤与验证

2.8.3.1　硬件部署

同 1.2.3.1 节。

2.8.3.2　工程调试

（1）将本项目的工程（12-Signal）文件夹复制到 RT-ThreadStudio\workspace 目录下。

（2）其余同 2.1.3.2 节。

2.8.3.3　验证效果

（1）关闭 RT-Thread Studio，拔掉仿真器，按下 ZI-ARMEmbed 上的电源按键重新上电。

（2）在 MobaXterm 串口终端 FinSH 控制台中输入"sig kill"命令，解析自定义 FinSH 控制台命令后，发送信号使 4 个 LED 依次点亮（见图 2.42），同时 FinSH 控制台显示"signal send success!"和信号值。

```
 \ | /
- RT -       Thread Operating System
 / | \       4.1.0 build Oct 14 2022 15:07:17
 2006 - 2022 Copyright by RT-Thread team
msh >sig kill
signal send success!
msh >thread received signal 30
```

图 2.42　通过发送信号实现 LED 流水灯效果

2.8.4　小结

本节主要介绍信号的基本概念、工作原理和管理方式。通过本节的学习，读者可掌握 RT-Thread 信号的应用。

2.9 RT-Thread 内存管理应用开发

内存管理负责有效地分配和释放内存资源，以及维护程序的内存空间。内存管理的目标是最大限度地提高系统的性能、可用性和安全性。本节主要介绍 RT-Thread 的两种内存管理方式，分别是动态内存管理和静态内存管理。

本节的要求如下：

- 了解内存管理的基本概念。
- 了解内存管理的工作原理。。
- 掌握内存的管理方式。
- 掌握 RT-Thread 内存管理接口的应用。

2.9.1　原理分析

2.9.1.1　内存管理

在计算机系统中，存储空间可以分为两种：内部存储空间和外部存储空间。内部存储空间（简称内存）的访问速度通常比较快，能够按照变量地址随机访问，就是我们常说的 RAM（随机存储器）；外部存储空间内所保存的数据在掉电后也不会丢失，就是我们常说的 ROM（只读存储器）。

在计算机系统中，变量、中间数据一般存放在 RAM 中，只有在实际使用时才将它们从 RAM 调入 CPU 中进行运算。这些数据需要的内存大小是在程序运行过程中根据实际情况来确定的，这就要求系统具有对内存进行动态管理的能力。在用户需要内存时，向系统申请，由系统选择一段合适的内存分配给用户，用户使用完毕后再释放内存，以便系统将该段内存回收再利用。

2.9.1.2　内存管理的功能特点

内存管理是操作系统中的一个重要组成部分，负责有效地分配和管理内存，以满足不同进程和应用程序的需求，同时确保内存的使用是安全的。内存管理包括以下几个关键方面：

（1）内存分配：内存管理器负责将计算机的物理内存划分为不同的区域，以便为进程分配内存。内存分配包括静态内存分配和动态内存分配。静态内存分配是在程序启动时分配内存的，动态内存分配是在程序运行中动态分配和释放内存的。

（2）虚拟内存：虚拟内存是一种将物理内存和磁盘空间组合在一起的技术，允许操作系统将进程的数据和代码分割成多个页面或块，以便在需要时将其加载到物理内存中。虚拟内存有助于提高内存的利用率和安全性。

（3）内存保护：内存保护负责确保不同进程之间的内存不会相互干扰，主要包括分配独立的内存空间、实施访问权限控制和检测内存溢出。

（4）内存回收：内存管理需要回收不再使用的内存，以便将其释放并分配给其他进程，内存回收主要包括释放被已经结束的进程占用的内存，以及检测和处理内存泄漏。

（5）碎片管理：内存管理还需要处理内存碎片问题。内部碎片是未被有效利用的内存块，碎片管理需要采取措施来减少内存碎片，提高内存的有效利用率。

（6）垃圾回收：对于程序运行中动态分配的内存，垃圾回收是一项重要工作，它负责检测和回收不再被程序使用的内存，以减少内存泄漏。

（7）内存交换和页面置换：如果物理内存不足，则可以使用内存交换和页面置换技术将部分进程或页面从内存中交换到磁盘上，以便腾出内存空间。

（8）内存分页和分段：可以将内存划分为页面或段，以便更加灵活地管理内存。分页和分段技术可以用于不同类型的内存管理。

在内存管理中，RT-Thread 根据上层应用及系统资源的不同，有针对性地提供了不同的内存管理算法。RT-Thread 的内存管理可分为内存堆管理与内存池管理。内存管理又可根据具体的内存分为 3 种情况：

（1）针对小内存块的分配管理（小内存管理算法）。

（2）针对大内存块的分配管理（slab 管理算法）。

（3）针对多内存堆的分配情况（memheap 管理算法）。

2.9.1.3　内存管理

内存管理用于管理一段连续的内存空间。RT-Thread 的内存堆分布情况如图 2.43 所示，RT-Thread 将"ZI 段结尾处"到内存尾部的空间用于内存堆。

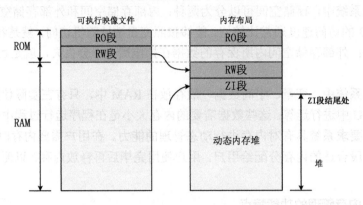

图 2.43　RT-Thread 的内存堆分布情况

在内存可以满足当前资源需求的情况下，RT-Thread 可根据用户对资源的需求分配任意大小的内存块。当用户无须再使用这些内存块时，又可以将这些内存块释放到内存中，以便供其他应用使用。为了满足不同的需求，RT-Thread 提供了不同的内存管理算法，分别是小内存管理算法、slab 管理算法和 memheap 管理算法。小内存管理算法主要针对系统资源比较少，一般用于需求小于 2 MB 内存空间的系统；而 slab 内存管理算法则在系统资源比较丰富的情况下提供了一种近似多内存池管理算法的快速算法。memheap 管理算法适用于系统存在多个内存堆的情况，它可以将多个内存堆"粘贴"在一起，构成一个大的内存堆。

2.9.1.4　内存的管理方式

内存管理的相关函数如图 2.44 所示，主要的操作包括内存配置和初始化、内存分配、内存释放。所有使用完成后的动态内存都应该被释放，以供其他应用使用。

图 2.44　内存管理的相关函数

1）内存配置和初始化

在使用内存时，必须要在系统初始化时进行内存配置和初始化。通过下面的函数可以完成内存的配置：

```
/******************************************************************************
 * 名称：rt_system_heap_init()
 * 功能：内存配置和初始化
 * 参数：begin_addr 表示内存的起始地址；end_addr 表示内存的结束地址
 ******************************************************************************/
void rt_system_heap_init(void*  begin_addr, void*  end_addr);
```

在使用 memheap 内存管理算法时，必须在系统初始化时对内存进行初始化。通过下面的函数可对内存进行初始化：

```
/*************************************************************************
 * 名称：rt_memheap_init()
 * 功能：内存配置和初始化
 * 参数：memheap 表示内存块；name 表示内存的名称；start_addr 表示内存的起始地址；size 表示内存的大小
 * 返回：RT_EOK 表示成功
 *************************************************************************/
rt_err_t rt_memheap_init(struct rt_memheap*    memheap, const char*    name, void*    start_addr,
rt_uint32_t    size)
```

如果有多个不连续的内存块，则可以多次调用 rt_memheap_init()函数将其初始化并加入 memheap_item 链表。

2）分配内存

通过下面的函数可分配内存：

```
/*************************************************************************
 * 名称：rt_malloc()
 * 功能：分配内存
 * 参数：nbytes 表示需要分配的内存块大小，单位为字节
 * 返回：分配的内存块地址表示成功；RT_NULL 表示失败
 *************************************************************************/
void *rt_malloc(rt_size_t    nbytes);
```

rt_malloc()函数会从系统内存空间中找到合适大小的内存块，然后把内存块可用地址返回给用户。

3）释放内存

在程序完成后，需要及时释放该程序使用的内存，否则会造成内存泄漏。通过下面的函数可释放内存：

```
/*************************************************************************
 * 名称：rt_free()
 * 功能：释放内存
 * 参数：ptr 表示待释放的内存块指针
 *************************************************************************/
void rt_free (void*    ptr);
```

rt_free()函数会把待释放的内存还给内存管理器，在调用该函数时需要传递待释放的内存块指针，如果是空指针则直接返回。

2.9.2　开发设计与实践

2.9.2.1　软件设计

软件设计流程如图 2.45 所示。

（1）在主函数中创建静态数组和动态数组，在这些数组中写入相同的数据，并显示数组

中的数据。

（2）在 MobaXterm 串口终端 FinSH 控制台中输入"memoryManage malloc 1"，FinSH 控制台输出"1 bytes of memory allocated success!"。

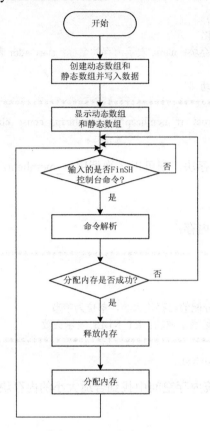

图 2.45　软件设计流程

2.9.2.2　功能设计与核心代码设计

通过原理和软件设计流程的学习，代码的设计就变得很清晰了。首先在主函数中创建静态数组和动态数组，然后通过自定义 FinSH 控制台命令来根据要求动态分配内存。

1）主函数（zonesion/app/main.c）

主函数的主要工作是创建静态数组和动态数组，并存储相同的数据。代码如下：

```
/*********************************************************************************
* 名称：main()
* 功能：创建动态数组和静态数组
*********************************************************************************/
#include <rtthread.h>
int main(void)
{
    //创建长度为 10 的静态数组和动态数组
    uint8_t array[10];
    uint8_t *parr = rt_malloc(10);                         //在两个数组中分别存储 0～9
    for(uint8_t i = 0; i < 10; i++) {
```

```
                array[i] = i;
                parr[i] = i;
        }
        for(uint8_t i = 0; i < 10; i++) {                        //显示两个数组中的数据
                rt_kprintf("array[%u]=%u\n", i, array[i]);
                rt_kprintf(" parr[%u]=%u\n", i, parr[i]);
        }
        rt_free(parr);                                          //释放动态分配的内存
        return 0;
}
```

2）自定义 FinSH 控制台命令函数（zonesion/app/finsh_cmd/finsh_ex.c）

自定义 FinSH 控制台命令函数的主要工作是根据要求动态分配内存。代码如下：

```
/*******************************************************************************
* 名称：memoryManage (int argc, char **argv)
* 功能：自定义 FinSH 控制台命令
* 参数：argc 表示参数个数；argv 表示指向参数字符串指针的数组指针
*******************************************************************************/
#include <rtthread.h>
#include <stdlib.h>
char *temp = RT_NULL;
int memoryManage(int argc, char **argv)
{
        if(!rt_strcmp(argv[1], "malloc"))
        {
                if(argc < 3)
                {
                        rt_kprintf("Please input malloc memory size!\n");
                        return 0;
                }
                int val = atoi(argv[2]);
                rt_free(temp);                                  //释放内存
                temp = RT_NULL;
                temp = rt_malloc(val);                          //分配内存
                if(temp != RT_NULL)
                        rt_kprintf("%d bytes of memory allocated success!\n", val);
                else
                        rt_kprintf("failed to allocate %d bytes of memory!\n", val);
        }
        else
                rt_kprintf("Please input 'memoryManage <malloc>'\n");
        return 0;
}
MSH_CMD_EXPORT(memoryManage, memory Manage sample);
```

2.9.3　开发步骤与验证

2.9.3.1　硬件部署

同 1.2.3.1 节。

2.9.3.2　工程调试

（1）将本项目的工程（13-Memory）文件夹复制到 RT-ThreadStudio\workspace 目录下。

（2）其余同 2.1.3.2 节。

2.9.3.3　验证效果

（1）关闭 RT-Thread Studio，拔掉仿真器，按下 ZI-ARMEmbed 上的电源按键重新上电。

（2）MobaXterm 串口终端 FinSH 控制台输出静态数组和动态数组中的数据。

```
 \ | /
- RT -     Thread Operating System
 / | \     4.1.0 build Oct 14 2022 15:15:09
 2006 - 2022 Copyright by RT-Thread team
array[0]=0
 parr[0]=0
array[1]=1
 parr[1]=1
array[2]=2
 parr[2]=2
array[3]=3
 parr[3]=3
array[4]=4
 parr[4]=4
array[5]=5
 parr[5]=5
array[6]=6
 parr[6]=6
array[7]=7
 parr[7]=7
array[8]=8
 parr[8]=8
array[9]=9
 parr[9]=9
```

（3）在 MobaXterm 串口终端 FinSH 控制台中输入“memoryManage malloc 1”命令，解析自定义 FinSH 控制台命令后可申请 1 B 的内存，FinSH 控制台输出“1 bytes of memory allocated success!”。

```
msh >memoryManage malloc
Please input malloc memory size!
msh >memoryManage malloc 1
1 bytes of memory allocated success!
```

```
msh >
```

（4）在 MobaXterm 串口终端 FinSH 控制台中输入"memoryManage malloc 999"命令，解析自定义 FinSH 控制台命令后可申请 999 B 的内存，FinSH 控制台输出"999 bytes of memory allocated success!"。

```
msh >memoryManage malloc 999
999 bytes of memory allocated success!
msh >
```

2.9.4　小结

本节主要介绍内存管理的基本概念、工作原理和管理方式。通过本节的学习，读者可掌握 RT-Thread 内存管理接口的应用。

第 3 章
RT-Thread 设备驱动开发技术

3.1 IO 设备驱动应用开发

IO（输入/输出）设备驱动程序是用于连接和管理外部设备与计算机系统或嵌入式系统之间通信的软件层，该驱动程序允许操作系统或应用程序与硬件设备进行通信，执行读写操作，并管理硬件设备的状态。

本节的要求如下：

◗ 了解 IO 设备的简介。
◗ 了解 IO 设备的分类。
◗ 掌握 IO 设备的管理方式。
◗ 掌握 RT-Thread IO 设备管理接口的应用。

3.1.1 原理分析

3.1.1.1 IO 设备的模型框架

绝大部分的嵌入式系统都包括一些 IO（Input/Output，输入/输出）设备，如仪器上的显示屏、工业设备上的串口通信、数据采集设备上用于保存数据的 Flash 或 SD 卡，以及网络设备的以太网接口等，这些都是嵌入式系统中常用的 IO 设备。

RT-Thread 提供了一套简单的 IO 设备模型框架，如图 3.1 所示。该框架位于硬件和应用程序之间，共分成三层，从上到下分别是 IO 设备管理层、IO 设备驱动框架层、IO 设备驱动层。

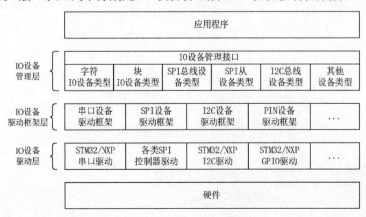

图 3.1　IO 设备的模型框架

（1）IO 设备管理层：实现了对 IO 设备驱动程序的封装。应用程序可通过 IO 设备管理层提供的标准接口访问底层 IO 设备，IO 设备驱动程序的升级、更替不会对上层应用产生影响。这种方式使得与 IO 设备硬件操作相关的代码能够独立于应用程序而存在，开发人员只需要关注各自的功能设计与核心代码设计，从而降低了代码的耦合性、复杂性，提高了系统的可靠性。

（2）IO 设备驱动框架层：对同类 IO 设备驱动进行抽象，将不同厂家的同类 IO 设备驱动中相同的部分抽取出来，为不同部分留出的接口由驱动程序实现。

（3）IO 设备驱动层：实际驱动 IO 设备工作的程序，实现访问 IO 设备的功能。

3.1.1.2　IO 设备的类型

RT-Thread 支持多种 IO 设备类型，主要的 IO 设备类型如表 3.1 所示。

表 3.1　RT-Thread 支持的主要 IO 设备类型

IO 设备类型	描　　述	IO 设备类型	描　　述
RT_Device_Class_Char	字符设备	RT_Device_Class_Block	块设备
RT_Device_Class_NetIf	网络接口设备	RT_Device_Class_MTD	内存设备
RT_Device_Class_RTC	RTC 设备	RT_Device_Class_Sound	声音设备
RT_Device_Class_Graphic	图形设备	RT_Device_Class_I2CBUS	I2C 设备
RT_Device_Class_USBDevice	USB 设备	RT_Device_Class_USBHost	USB 主设备
RT_Device_Class_SPIBUS	SPI 总线设备	RT_Device_Class_SPIDevice	SPI 设备
RT_Device_Class_SDIO	SDIO 设备	RT_Device_Class_Miscellaneous	其他设备

其中字符设备、块设备是常用的 IO 设备类型，它们的分类依据是 IO 设备数据与系统之间的传输处理方式。字符设备支持非结构型数据的传输，即数据传输采用串行的形式，每次传输 1 B 的数据。字符设备通常是一些简单 IO 设备，如串口、按键。块设备每次传输 1 个数据块，如每次传输 512 B 的数据。这个数据块是硬件强制性的，数据块可能使用某类数据接口或某些强制性的传输协议，否则就可能发生错误。

3.1.1.3　IO 设备的管理方式

1）创建 IO 设备

IO 设备驱动层负责创建 IO 设备，并注册到 IO 设备管理器中，用户既可以通过静态申明的方式创建 IO 设备，也可以用下面的函数动态创建 IO 设备。

```
/*************************************************************************
 * 名称：rt_device_create()
 * 功能：创建 IO 设备
 * 参数：type 表示 IO 设备类型；attach_size 表示用户数据大小
 * 返回：IO 设备句柄表示创建成功；RT_NULL 表示创建失败
 *************************************************************************/
rt_device_t rt_device_create(int type, int attach_size);
```

调用 rt_device_create()函数时，RT-Thread 会从动态地从内存中分配一个 IO 设备控制块，大小为 struct rt_device 和 attach_size 的和，IO 设备的类型由参数 type 设定。IO 设备被创建后，还需要实现访问该 IO 设备的操作方法。

```
struct rt_device_ops
{
    //common device interface
    rt_err_t    (*init)    (rt_device_t dev);                                         //初始化 IO 设备
    rt_err_t    (*open)    (rt_device_t dev, rt_uint16_t oflags);                      //打开 IO 设备
    rt_err_t    (*close)   (rt_device_t dev);                                         //关闭 IO 设备
    rt_size_t (*read)     (rt_device_t dev, rt_off_t pos, void *buffer, rt_size_t size);      //从 IO 设备读取数据
    rt_size_t (*write)    (rt_device_t dev, rt_off_t pos, const void *buffer, rt_size_t size);//向 IO 设备写入数据
    rt_err_t    (*control)(rt_device_t dev, int cmd, void *args);                     //控制 IO 设备
};
```

2）销毁 IO 设备

当不再使用动态创建的 IO 设备时，可通过下面的函数来删除该 IO 设备：

```
/*************************************************************************
 * 名称：rt_device_destroy()
 * 功能：销毁 IO 设备
 * 参数：device 表示要销毁的 IO 设备句柄
 *************************************************************************/
void rt_device_destroy(rt_device_t device);
```

3）注册 IO 设备

IO 设备被创建后，需要将其注册到 IO 设备管理器中，这样应用程序才能够访问该 IO 设备。通过下面的函数可注册 IO 设备：

```
/*************************************************************************
 * 名称：rt_device_register()
 * 功能：注册 IO 设备
 * 参数：dev 表示 IO 设备句柄；name 表示 IO 设备名称；flags 表示 IO 设备模式标志
 * 返回：RT_EOK 表示注册成功，-RT_ERROR 表示注册失败
 *************************************************************************/
rt_err_t rt_device_register(rt_device_t dev, const char* name, rt_uint8_t flags);
```

成功注册 IO 设备后，可以在 FinSH 控制台中使用"list_device"命令查看系统中所有的 IO 设备信息，包括 IO 设备名称、IO 设备类型和 IO 设备被打开的次数。

4）注销 IO 设备

当 IO 设备注销后的，IO 设备将从 IO 设备管理器中删除，也就不能再通过查找 IO 设备函数来搜索到该 IO 设备。注销 IO 设备不会释放 IO 设备控制块占用的内存。注销 IO 设备的函数如下：

```
/*************************************************************************
 * 名称：rt_device_unregister()
 * 功能：注销 IO 设备
 * 参数：dev 表示 IO 设备句柄
 * 返回：RT_EOK 表示注销成功；-RT_ERROR 表示注销失败
 *************************************************************************/
rt_err_t rt_device_unregister(rt_device_t dev);
```

5）查找 IO 设备

应用程序可根据 IO 设备的名称来获取 IO 设备句柄，进而操作该 IO 设备。查找 IO 设备函数如下：

```
/**********************************************************************************
 * 名称：rt_device_find()
 * 功能：查找 IO 设备
 * 参数：name 表示 IO 设备名称
 * 返回：IO 设备句柄表示查找到对应的 IO 设备；RT_NULL 表示没有找到相应的 IO 设备
 **********************************************************************************/
rt_device_t rt_device_find(const char* name);
```

6）初始化 IO 设备

获得 IO 设备句柄后，应用程序可使用下面的函数对 IO 设备进行初始化操作：

```
/**********************************************************************************
 * 名称：rt_device_init()
 * 功能：初始化 IO 设备
 * 参数：dev 表示 IO 设备句柄
 * 返回：RT_EOK 表示 IO 设备初始化成功；错误码表示 IO 设备初始化失败
 **********************************************************************************/
rt_err_t rt_device_init(rt_device_t dev);
```

如果调用 rt_device_init()函数完成了 IO 设备的初始化，则再次调用该函数时将不再重复进行初始化。

7）打开 IO 设备

通过 IO 设备句柄，应用程序可以打开和关闭 IO 设备，打开 IO 设备时，会先检测 IO 设备是否已经初始化，如果没有初始化则会默认调用 rt_device_init()函数进行初始化。通过下面的函数可打开 IO 设备：

```
/**********************************************************************************
 * 名称：rt_device_open()
 * 功能：打开 IO 设备
 * 参数：dev 表示 IO 设备句柄
 * 参数：oflags 表示打开 IO 设备的模式标志
 * 返回：RT_EOK 表示成功打开 IO 设备；-RT_EBUSY 表示 IO 设备不允许重复打开；其他错误码表
示打开失败
 **********************************************************************************/
rt_err_t rt_device_open(rt_device_t dev, rt_uint16_t oflags);
```

8）关闭 IO 设备

在完成对 IO 设备的操作后，如果不再需要对该 IO 设备进行操作，则可以关闭该 IO 设备。通过下面的函数可以关闭 IO 设备：

```
/**********************************************************************************
 * 名称：rt_device_close()
 * 功能：关闭 IO 设备
 * 参数：dev 表示 IO 设备句柄
 * 返回：RT_EOK 表示成功关闭 IO 设备；-RT_ERROR 表示 IO 设备不可重复关闭；其他错误码表示
```

关闭失败
```
*********************************************************************************/
rt_err_t rt_device_close(rt_device_t dev);
```

9）控制 IO 设备

通过命令控制字，应用程序也可以控制 IO 设备。通过下面的函数可控制 IO 设备：

```
/*********************************************************************************
 * 名称：rt_device_control()
 * 功能：控制 IO 设备
 * 参数：dev 表示 IO 设备句柄；cmd 表示命令控制字；arg 表示控制的参数
 * 返回：RT_EOK 表示执行成功；-RT_ENOSYS 表示 dev 为空，执行失败；其他错误码表示执行失败
 *********************************************************************************/
rt_err_t rt_device_control(rt_device_t dev, rt_uint8_t cmd, void* arg);
```

3.1.2　开发设计与实践

3.1.2.1　硬件设计

本节的硬件设计同 1.5.2.1 节。本节通过 MobaXterm 串口终端 FinSH 控制台实现对 LED 的控制，目的是帮助读者掌握 IO 设备的管理方式。

3.1.2.2　软件设计

软件设计流程如图 3.2 所示。

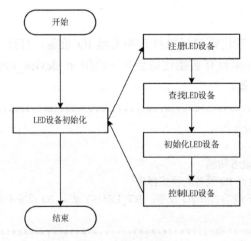

图 3.2　软件设计流程

3.1.2.3　功能设计与核心代码设计

通过原理分析可知，要实现对 LED 设备的管理，首先需要使用 RT-Thread 提供的设备管理接口初始化 LED 设备，然后实现对 LED 设备的各种操作。

1）主函数（zonesion/app/main.c）

主函数的主要工作是完成 LED 设备的初始化，初始化完成后退出主函数。代码如下：

```
/*********************************************************************************
 * 名称：main()
```

```
* 功能：LED 设备初始化
**********************************************************************************/
#include "devModel.h"
int main(void)
{
    ledDev_init();                                          //LED 设备初始化
    return 0;
}
```

2）LED 设备初始化函数（zonesion/app/devModel.c）

LED 设备初始化函数的主要工作是实现设备的注册、查找、初始化和控制。代码如下：

```
/*********************************************************************************
* 名称：ledDev_init()
* 功能：LED 初始化
**********************************************************************************/
int ledDev_init(void)
{
    rt_err_t result = RT_EOK;
    result = rt_hw_led_register(LED_DEV_NAME, RT_NULL);     //注册 LED 设备
    if(result != RT_EOK)                                    //判断是否注册成功
    {
        rt_kprintf("failed to register %s device!\n", LED_DEV_NAME);
        return -1;
    }
    ledDev = rt_device_find(LED_DEV_NAME);                  //查找 LED 设备
    if(ledDev == RT_NULL)                                   //判断是否查找到设备
    {
        rt_kprintf("failed to find %s device!\n", LED_DEV_NAME);
        return -1;
    }
    result = rt_device_init(ledDev);                        //初始化 LED 设备
    if(result != RT_EOK)                                    //判断是否初始化成功
    {
        rt_kprintf("failed to init %s device!\n", LED_DEV_NAME);
        return -1;
    }
    unsigned char state = 0x0A;
    result = rt_device_control(ledDev, 0, &state);          //设置 LED 设备状态
    if(result != RT_EOK)                                    //判断是否设置状态成功
    {
        rt_kprintf("failed to control %s device!\n", LED_DEV_NAME);
        return -1;
    }
    return 0;
}
```

3）LED 设备注册函数（zonesion/app/devModel.c）

LED 设备注册函数的主要工作是完成 LED 设备的创建，设备的创建的方式可以选择动

态创建或者静态创建。代码如下：

```
/*****************************************************************************
 * 名称：rt_hw_led_register()
 * 功能：LED 设备注册
 * 参数：name 表示设备名称；user_data 表示用户数据，可选 RT_NULL
 * 返回：RT_EOK 表示注册成功；-RT_ERROR 表示注册失败
 *****************************************************************************/
static rt_err_t rt_hw_led_register(const char *name, void *user_data)
{
    device.type = RT_Device_Class_Miscellaneous;                //设备类型：其他设备
    device.rx_indicate = RT_NULL;                               //接收回调
    device.tx_complete = RT_NULL;                               //发送完成
    device.init = led_init;                                     //设备初始化
    device.open = RT_NULL;                                      //设备打开
    device.close = RT_NULL;                                     //设备关闭
    device.read = RT_NULL;                                      //设备读取
    device.write = RT_NULL;                                     //设备写入
    device.control = led_ctrlState;                            //设备控制
    device.user_data = user_data;                              //设备用户数据：无
    return rt_device_register(&device, name, RT_DEVICE_FLAG_RDWR); //注册驱动
}
```

4）LED 设备初始化函数（zonesion/app/devModel.c）

LED 设备初始化函数的主要工作是完成 LED 设备 GPIO 的初始化，包括设置输出模式和初始状态。代码如下：

```
/*****************************************************************************
 * 名称：led_init()
 * 功能：LED 设备初始化接口
 * 参数：dev 表示设备操作句柄
 *****************************************************************************/
static rt_err_t led_init(rt_device_t dev)
{
    led_pin_init();                                             //初始化 LED 设备的 GPIO
    return 0;
}
```

实际的 GPIO 初始化是在 led_pin_init 函数（zonesion/common/DRV/drv_led/drv_led.c）中完成的。代码如下：

```
/*****************************************************************************
 * 名称：led_pin_init()
 * 功能：LED 设备 GPIO 的初始化
 *****************************************************************************/
void led_pin_init(void)
{
    rt_pin_mode(LED1_PIN_NUM, PIN_MODE_OUTPUT);                //将 LED1 的 GPIO 设置为输出模式
    rt_pin_mode(LED2_PIN_NUM, PIN_MODE_OUTPUT);                //将 LED2 的 GPIO 设置为输出模式
```

```
rt_pin_mode(LED3_PIN_NUM, PIN_MODE_OUTPUT);          //将 LED3 的 GPIO 设置为输出模式
rt_pin_mode(LED4_PIN_NUM, PIN_MODE_OUTPUT);          //将 LED4 的 GPIO 设置为输出模式
rt_pin_write(LED1_PIN_NUM, PIN_HIGH);                //将 LED1 的 GPIO 设置为高电平
rt_pin_write(LED2_PIN_NUM, PIN_HIGH);                //将 LED2 的 GPIO 设置为高电平
rt_pin_write(LED3_PIN_NUM, PIN_HIGH);                //将 LED3 的 GPIO 设置为高电平
rt_pin_write(LED4_PIN_NUM, PIN_HIGH);                //将 LED4 的 GPIO 设置为高电平
}
```

5）LED 设备控制函数（zonesion/common/DRV/drv_led/drv_led.c）

LED 设备控制函数的主要工作是完成对不同 LED 的控制，通过传入的整型 cmd 命令按位控制 LED。代码如下：

```
/*********************************************************************
* 名称：led_ctrl()
* 功能：LED 控制
* 参数：cmd 表示控制值
*********************************************************************/
void led_ctrl(unsigned char cmd)
{
    rt_pin_write(LED1_PIN_NUM, !(cmd & LED1_NUM));
    rt_pin_write(LED2_PIN_NUM, !(cmd & LED2_NUM));
    rt_pin_write(LED3_PIN_NUM, !(cmd & LED3_NUM));
    rt_pin_write(LED4_PIN_NUM, !(cmd & LED4_NUM));
}
```

6）自定义 FinSH 控制台命令函数（zonesion/app/finsh_cmd/finsh_ex.c）

自定义 FinSH 控制台命令函数的主要工作是实现 IO 设备管理命令，通过命令控制 4 个 LED。代码如下：

```
/*********************************************************************
* 名称：devControl()
* 功能：自定义 FinSH 控制台命令
* 参数：argc 表示参数个数；argv 表示指向参数字符串指针的数组指针
*********************************************************************/
int devControl(int argc, char **argv)
{
    if(argc < 2)
    {
        rt_kprintf("Please input device status!\n");
        return 0;
    }
    int status = strtol(argv[1],0,0);
    rt_err_t result = rt_device_control(ledDev, 0, &status);       //设置 LED 设备状态
    if(result != RT_EOK)                                           //判断是否成功设置状态
    {
        rt_kprintf("failed to control led device!\n");
        return -1;
    }
    return 0;
```

```
}
MSH_CMD_EXPORT(devControl, device control sample);
```

3.1.3 开发步骤与验证

3.1.3.1 硬件部署

同 1.2.3.1 节。

3.1.3.2 工程调试

（1）将本项目的工程（14-DeviceModel）文件夹复制到 RT-ThreadStudio\workspace 目录下。
（2）其余同 2.1.3.2 节。

3.1.3.3 验证效果

（1）关闭 RT-Thread Studio，拔掉仿真器，按下 ZI-ARMEmbed 上的电源按键重新上电。
（2）在 MobaXterm 串口终端 FinSH 控制台中输入命令"devControl 1"，解析自定义 FinSH 控制台命令后点亮 LED1（见图 3.3）。

```
 \ | /
- RT -      Thread Operating System
 / | \      4.1.0 build Oct 20 2022 16:34:36
 2006 - 2022 Copyright by RT-Thread team
msh >devControl 1
```

（3）在 MobaXterm 串口终端 FinSH 控制台中输入命令"devControl 3"，解析自定义 FinSH 控制台命令后同时点亮 LED1 和 LED2（见图 3.4）。

```
 \ | /
- RT -      Thread Operating System
 / | \      4.1.0 build Oct 20 2022 16:34:36
 2006 - 2022 Copyright by RT-Thread team
msh >devControl 1
msh >devControl 3
```

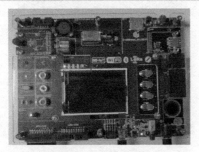

图 3.3 点亮 LED1

图 3.4 点亮 LED1 和 LED2

3.1.4 小结

本节主要介绍 IO 设备的模型框架、类型和管理方式。通过本节的学习，读者可掌握 RT-Thread IO 设备管理接口的应用。

3.2 UART 设备驱动应用开发

串口是一种用于串行通信的物理接口和通信协议。它是一种在计算机和外部设备之间传输数据的标准接口。串口通信通常以串行方式传输数据，相对于并行通信来说，只使用一条传输线。常见的串口标准包括 RS-232、RS-485、USB 等。

本节的要求如下：

- 了解 UART 设备的工作原理与基本概念。
- 掌握 UART 设备的管理方式。
- 掌握 RT-Thread UART 设备管理接口的应用。

3.2.1　原理分析

3.2.1.1　UART 简介

通用异步接收发送设备（Universal Asynchronous Receiver/Transmitter，UART）是一种异步串口通信协议，采用逐位传输的方式，在应用程序开发过程中的使用频率非常高。

串行通信的特点是数据按位传输，最少只需一根传输线即可完成；成本低但传输速率低。

串口通信常用的参数有波特率、数据位、停止位和奇偶校验位，通信双方的参数必须匹配。UART 协议的数据帧如图 3.5 所示。

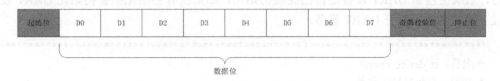

图 3.5　UART 协议的数据帧

（1）起始位：表示数据传输的开始，逻辑电平为"0"。

（2）数据位：可能值有 5、6、7、8、9，表示传输的数据位数。一般取值为 8，因为 ASCII 码中的字符是 8 位的。

（3）奇偶校验位：用于接收方对接收到的数据进行校验，校验"1"的位数为偶数（偶校验）或奇数（奇校验），以此来校验数据传输的正确性，使用时不需要此位也可以。

（4）停止位：表示一帧数据的结束，逻辑电平为"1"。

（5）波特率：串口通信时的速率，用单位时间内传输的二进制代码的有效位（bit）来表示，其单位为每秒比特（bit/s、bps）。常见的波特率值有 4800、9600、14400、38400、115200 等，数值越大表示数据传输速率越快，波特率为 115200 表示每秒传输 115200 bit 的数据。

3.2.1.2　UART 设备的管理方式

UART 设备管理接口如表 3.2 所示。

表 3.2　UART 设备管理接口

UART 设备管理接口	描　述	UART 设备管理接口	描　述
rt_device_find()	查找设备	rt_device_open()	打开设备
rt_device_read()	读取数据	rt_device_write()	写入数据
rt_device_control()	控制设备	rt_device_set_rx_indicate()	设置接收回调函数
rt_device_set_tx_complete()	设置发送完成回调函数	rt_device_close()	关闭设备

1）查找 UART 设备

应用程序根据 UART 设备名称获取 UART 设备句柄后，可以操作该 UART 设备。通过下面的函数可以查找 UART 设备：

```
/********************************************************************************
 * 名称：rt_device_find()
 * 功能：查找 UART 设备
 * 参数：name 表示 UART 设备名称
 * 返回：UART 设备句柄表示查找到对应 UART 设备；RT_NULL 表示没有找到相应的 UART 设备
 ********************************************************************************/
rt_device_t rt_device_find(const char* name);
```

2）打开 UART 设备

通过 UART 设备句柄，应用程序可以打开和关闭对应的 UART 设备。在打开 UART 设备时，系统会检测 UART 设备是否已经被初始化，如果没有初始化则会初始化 UART 设备。通过下面的函数可打开 UART 设备：

```
/********************************************************************************
 * 名称：rt_device_open()
 * 功能：打开 UART 设备
 * 参数：dev 表示 UART 设备句柄；oflags 表示打开 UART 设备的模式
 * 返回：RT_EOK 表示 UART 设备打开成功；-RT_EBUSY 表示 UART 设备不允许重复打开；其他
错误码表示打开失败
 ********************************************************************************/
rt_err_t rt_device_open(rt_device_t dev, rt_uint16_t oflags);
```

3）关闭 UART 设备

当应用程序完成对 UART 设备的操作后，如果不需要再进行操作，则可以关闭 UART 设备。通过下面的函数可关闭 UART 设备：

```
/********************************************************************************
 * 名称：rt_device_close()
 * 功能：关闭 UART 设备
 * 参数：dev 表示 UART 设备句柄
 * 返回：RT_EOK 表示 UART 设备关闭成功；-RT_ERROR 表示 UART 设备不可重复关闭；其他错
误码表示关闭失败
 ********************************************************************************/
rt_err_t rt_device_close(rt_device_t dev);
```

4）控制 UART 设备

通过命令控制字，应用程序可以控制 UART 设备。通过下面的函数可控制 UART 设备：

```
/******************************************************************************
 * 名称：rt_device_control()
 * 功能：控制 UART 设备
 * 参数：dev 表示 UART 设备句柄；cmd 表示命令控制字，其值为 RT_DEVICE_CTRL_CONFIG；arg
表示控制的参数，其类型：struct serial_configure
 * 返回：RT_EOK 表示执行成功；-RT_ENOSYS 表示 dev 为空，执行失败；其他错误码表示执行失败
 ******************************************************************************/
rt_err_t rt_device_control(rt_device_t dev, rt_uint8_t cmd, void* arg);
```

5）发送数据

通过下面的函数可以向 UART 设备发送数据（写入数据）：

```
/******************************************************************************
 * 名称：rt_device_write()
 * 功能：发送数据
 * 参数：dev 表示 UART 设备句柄；pos 表示写入数据的偏移量，此参数在 UART 设备中未使用；buffer
表示指向缓冲区的指针，放置要写入的数据；size 表示写入数据的大小
 * 返回：写入数据的实际大小表示写入数据成功；0 表示需要读取当前线程的 errno 来判断错误状态
 ******************************************************************************/
rt_size_t rt_device_write(rt_device_t dev, rt_off_t pos, const void* buffer, rt_size_t size);
```

6）接收数据

通过下面的函数可以向 UART 设备接收数据（读取数据）：

```
/******************************************************************************
 * 名称：rt_device_read()
 * 功能：接收数据
 * 参数：dev 表示 UART 设备句柄；pos 表示读取数据偏移量，此参数在 UART 设备中未使用；buffer
表示指向缓冲区的指针，读取的数据将会保存在缓冲区中；size 表示读取数据的大小
 * 返回：读取数据的实际大小表示读取数据成功；0 表示需要读取当前线程的 errno 来判断错误状态
 ******************************************************************************/
rt_size_t rt_device_read(rt_device_t dev, rt_off_t pos, void* buffer, rt_size_t size);
```

3.2.2　开发设计与实践

3.2.2.1　硬件设计

本节将通过 MobaXterm 串口终端 FinSH 控制台实现对信号灯的控制，目的是帮助读者掌握 UART 设备管理。FinSH 控制台和 UART 的连接如图 3.6 所示。

实现两个 UART 设备的通信需要四根信号线，分别是 VCC、GND、TX、RX，其中 VCC 连接电源，GND 用于接地，TX 和 RX 分别用于发送数据和接收数据。如果两个 UART 设备使用的逻辑电平标准相同，则可以直接连接。但在实际的应用场景中，经常需要将 UART 设备与计算机连接，UART 设备使用的是 TLL 电平，计算机使用的是 RS-232 电平，要实现两者的正常通信需要进行电平转换。电平转换电路如图 3.7 所示。

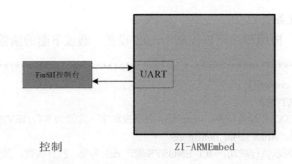

图 3.6 FinSH 控制台和 UART 的连接

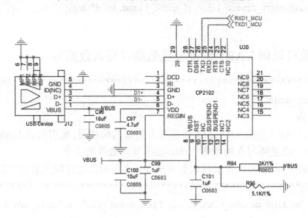

图 3.7 电平转换电路

3.2.2.2 软件设计

软件设计流程如图 3.8 所示。

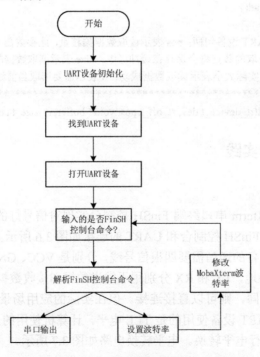

图 3.8 软件设计流程

（1）初始化 UART 设备。

（2）在主线程中找到 UART 设备并打开 UART 设备。

（3）在 MobaXterm 串口终端 FinSH 控制台中输入命令来控制 UART 设备。

3.2.2.3　功能设计与核心代码设计

要控制 UART 设备，首先需要使用 RT-Thread 提供的 UART 设备管理接口进行 UART 设备的初始化，然后进行相应的操作，如查找、打开等。

1）主函数（zonesion/app/main.c）

主函数的主要工作是查找并打开 UART 设备，完成之后退出主函数。代码如下：

```
/************************************************************************
* 名称：main()
* 功能：查找并打开 UART 设备
************************************************************************/
int main(void)
{
    uartDev = rt_device_find(RT_CONSOLE_DEVICE_NAME);          //查找 UART 设备
    if(uartDev == RT_NULL)                                     //判断是否找到 UART 设备
    {
        rt_kprintf("failed to find %s device!\n", RT_CONSOLE_DEVICE_NAME);
        return -1;
    }
    //此处只为演示 UART 设备的使用方法，实际上在 FinSH 控制台初始化时已经将 UART 设备打开
    rt_device_open(uartDev, RT_DEVICE_OFLAG_RDWR | RT_DEVICE_FLAG_STREAM);
    return 0;
}
```

2）自定义 FinSH 控制台命令函数模块（zonesion\app\finsh_cmd\finsh_ex.c）

自定义 FinSH 控制台命令函数的主要工作是实现控制 UART 设备的命令，包括输出串口信息、设置波特率，可以在 FinSH 控制台中使用相关命令对 UART 设备进行控制。代码如下：

```
/************************************************************************
* 名称：uartDrv()
* 功能：自定义 FinSH 控制台命令
* 参数：argc 表示参数个数；argv 表示指向参数字符串指针的数组指针
************************************************************************/
#include <rtthread.h>
#include <string.h>
#include <stdlib.h>
#include "rtdevice.h"
extern rt_device_t uartDev;
int uartDrv(int argc, char **argv)
{
    if(!rt_strcmp(argv[1], "show"))
    {
        if(argc < 3)
```

```
        {
            rt_kprintf("Please enter serial data!\n");
            return 0;
        }
        rt_device_write(uartDev, 0, argv[2], strlen(argv[2]));              //写入 UART 设备
        rt_device_write(uartDev, 0, "\r\n", strlen("\r\n"));                //换行显示
    } else if(!rt_strcmp(argv[1], "baudrate")) {
        struct serial_configure param = RT_SERIAL_CONFIG_DEFAULT;
        uint32_t baud = atoi(argv[2]);
        if(baud!=BAUD_RATE_2400 && baud!=BAUD_RATE_4800 && baud!=BAUD_RATE_9600
            && baud!=BAUD_RATE_19200 && baud!=BAUD_RATE_38400
            && baud!=BAUD_RATE_57600
            && baud!=BAUD_RATE_115200 && baud!=BAUD_RATE_230400
            && baud!=BAUD_RATE_460800 \
            && baud!=BAUD_RATE_921600 && baud!=BAUD_RATE_2000000
            && baud!= BAUD_RATE_3000000) {
            rt_kprintf("Illegal Baudrate!\n");
            return 0;
        }
        rt_kprintf("Baudrate set to %u\n", baud);
        param.baud_rate = baud;
        rt_device_control(uartDev, RT_DEVICE_CTRL_CONFIG, &param);
    }
    else
        rt_kprintf("Please input 'uartDrv <ut>'\n");
    return 0;
}
MSH_CMD_EXPORT(uartDrv, uart driver sample);
```

3.2.3　开发步骤与验证

3.2.3.1　硬件部署

同 1.2.3.1 节。

3.2.3.2　工程调试

（1）将本项目的工程（15-UART）文件夹复制到 RT-ThreadStudio\workspace 目录下。

（2）其余同 2.1.3.2 节。

3.2.3.3　验证效果

（1）关闭 RT-Thread Studio，拔掉仿真器，按下 ZI-ARMEmbed 上的电源按键重新上电。

（2）在 MobaXterm 串口终端 FinSH 控制台中输入 "uartDrv show helloworld" 命令，解析自定义 FinSH 控制台命令后在主线程中找到并打开 UART 设备，FinSH 控制台输出 "helloworld"。

```
 \|/
- RT -      Thread Operating System
 /|\      4.1.0 build Oct 20 2022 16:57:16
 2006 - 2022 Copyright by RT-Thread team
msh >uartDrv show helloworld
helloworld
msh >
```

（3）在 MobaXterm 串口终端 FinSH 控制台中输入 "uartDrv baudrate 1234" 命令，解析自定义 FinSH 控制台命令后，设置串口波特率为 1234，FinSH 控制台输出 "Illegal Baudrate!"。

```
 \|/
- RT -      Thread Operating System
 /|\      4.1.0 build Oct 20 2022 16:57:16
 2006 - 2022 Copyright by RT-Thread team
msh >uartDrv show helloworld
helloworld
msh >uartDrv baudrate 1234
Illegal Baudrate!
```

（4）在 MobaXterm 串口终端 FinSH 控制台中输入 "uartDrv baudrate 9600" 命令，解析自定义 FinSH 控制台命令后，将串口波特率修改为 9600，FinSH 控制台打印 "Baudrate set to 9600"。

```
 \|/
- RT -      Thread Operating System
 /|\      4.1.0 build Oct 20 2022 16:57:16
 2006 - 2022 Copyright by RT-Thread team
msh >uartDrv show helloworld
helloworld
msh >uartDrv baudrate 1234
Illegal Baudrate!
msh >uartDrv baudrate 9600
Baudrate set to 9600
```

3.2.4　小结

本节主要介绍 UART 设备的基本概念和管理方式。通过本节的学习，读者可掌握 RT-Thread UART 设备管理接口的应用。

3.3 PIN 设备驱动应用开发

在计算机系统和嵌入式系统中，PIN 设备通常是指可编程的输入/输出引脚（Programmable Input/Output Pins）。这些引脚可以通过软件进行配置，用于连接和控制外部设备，如传感器、执行器、LED 等。

本节的要求如下：

⊃ 了解 PIN 设备的基本概念。

⊃ 掌握 PIN 设备的管理方式。

⊃ 掌握 RT-Thread PIN 设备管理接口的应用。

3.3.1　原理分析

3.3.1.1　PIN 设备简介

MCU 上的引脚一般分为 4 类：电源、时钟、控制与 IO。IO 在使用模式上又可分为通用输入/输出（General Purpose Input Output，GPIO）和功能复用 IO（如 SPI、I2C、UART 等）。

大多数 MCU 的引脚都不止一个功能。不同引脚的内部结构不同，具有的功能也不一样。通过不同的配置可切换引脚的功能。GPIO 主要特性如下：

（1）可编程控制中断：中断触发模式可配置，一般有如图 3.9 所示的 5 种中断触发模式。

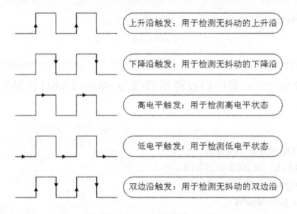

图 3.9　5 种中断触发模式

（2）输入/输出模式可控制。输出模式一般包括推挽、开漏、上拉、下拉，当引脚设置为输出模式时，可以通过将引脚输出的电平状态配置为高电平或低电平来控制外部设备。输入模式一般包括浮空、上拉、下拉、模拟，当引脚设置为输入模式时，可以读取引脚的电平状态，即高电平或低电平。

PIN 设备的基本信息如下：

（1）GPIO 引脚。

⊃ 通用性：GPIO 引脚是通用的输入/输出引脚，可以在需要的时候通过软件配置为输入引脚或输出引脚。

⊃ 数字信号：GPIO 引脚通常用于传输数字信号，即高电平和低电平。

（2）应用场合。

⊃ 传感器接口：PIN 设备可用于连接各种传感器，如温度传感器、湿度传感器、光敏传感器等。

⊃ 控制执行器：PIN 设备可以用于控制执行器，如电机、继电器等。

⊃ 用户界面：在某些嵌入式系统中，PIN 设备可以用于连接按钮、开关、LED 等，用于实现用户界面。

（3）编程和配置。

⊃ 寄存器设置：配置和控制 PIN 设备需要通过寄存器设置，通过特定的寄存器，可以配置 PIN 设备的输入/输出状态和电平。

⊃ 驱动程序：操作系统或嵌入式系统通常提供相应的驱动程序或 API，以便开发人员能方便地配置和使用 PIN 设备。

3.3.1.2　PIN 设备管理方式

1）设置引脚模式

在使用 PIN 设备前，需要先设置引脚模式。通过下面的函数可设置引脚模式：

```
/*****************************************************************************
 * 名称：rt_pin_mode()
 * 功能：设置引脚模式
 * 参数：pin 表示引脚编号；mode 表示引脚模式
 *****************************************************************************/
void rt_pin_mode(rt_base_t pin, rt_base_t mode);
```

2）设置引脚电平

通过下面的函数可设置引脚电平：

```
/*****************************************************************************
 * 名称：rt_pin_write()
 * 功能：设置引脚电平
 * 参数：pin 表示引脚编号；value 表示电平逻辑值，其值为 PIN_LOW（低电平）或 PIN_HIGH（高
电平）
 *****************************************************************************/
void rt_pin_write(rt_base_t pin, rt_base_t value);
```

3）读取引脚电平

通过下面的函数可读取引脚电平：

```
/*****************************************************************************
 * 名称：rt_pin_read()
 * 功能：读取引脚电平
 * 参数：pin 表示引脚编号
 * 返回：PIN_LOW 表示低电平；PIN_HIG 表示高电平
 *****************************************************************************/
int rt_pin_read(rt_base_t pin);
```

4）绑定引脚中断回调函数

在需要使用引脚的中断功能时，可通过下面的函数将某个引脚配置为某种中断触发模式并绑定一个中断回调函数。当引脚发生中断时，就会执行中断回调函数。

```
/*****************************************************************************
 * 名称：rt_pin_attach_irq()
 * 功能：绑定引脚中断回调函数
 * 参数：pin 表示引脚编号；mode 表示中断触发模式；hdr 表示中断回调函数，用户需要自行定义该
函数；args 表示中断回调函数的参数，不需要时设置为 RT_NULL
```

```
 * 返回：RT_EOK 表示绑定成功；错误码表示绑定失败
 ********************************************************************************/
rt_err_t rt_pin_attach_irq(rt_int32_t pin, rt_uint32_t mode, void (*hdr)(void *args), void *args);
```

5）使能引脚中断

通过下面的函数可使能引脚中断：

```
/********************************************************************************
 * 名称：rt_pin_irq_enable()
 * 功能：使能引脚中断
 * 参数：pin 表示引脚编号；enabled 表示状态，其值为 PIN_IRQ_ENABLE（开启）或 PIN_IRQ_DISABLE
（关闭）
 * 返回：RT_EOK 表示使能成功；错误码表示使能失败
 ********************************************************************************/
rt_err_t rt_pin_irq_enable(rt_base_t pin, rt_uint32_t enabled);
```

6）脱离（解绑定）引脚中断回调函数

通过下面的函数可以脱离引脚中断回调函数：

```
/********************************************************************************
 * 名称：rt_pin_detach_irq()
 * 功能：脱离引脚中断回调函数
 * 参数：pin 表示引脚编号
 * 返回：RT_EOK 表示脱离成功；错误码表示脱离失败
 ********************************************************************************/
rt_err_t rt_pin_detach_irq(rt_int32_t pin);
```

3.3.2　开发设计与实践

3.3.2.1　硬件设计

本节通过按键来控制 RGB 灯，以帮助读者理解 PIN 设备的管理方法。按键（KEY）、RGB 灯、ZI-ARMEmbed 和 FinSH 控制台的连接如图 3.10 所示。

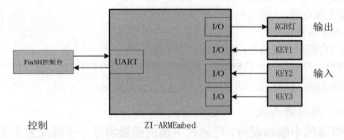

图 3.10　按键、RGB 灯、ZI-ARMEmbed 和 FinSH 控制台

RGB 灯与 ZI-ARMEmbed 的连接如图 3.11 所示，RGB 灯的一端通过电阻连接在 ZI-ARMEmbed 的 STM32F407 上，另一端连接在 3.3 V 的电源上。RGB 灯采用的是正向导通的方式，当 PB0、PB1 和 PB2 为高电平（3.3 V）时，RGB 灯（RGB_R、RGB_G 和 RGB_B）两端电压相同，无法形成压降，RGB 灯熄灭。反之，当 PB0、PB1 和 PB2 为低电平时，RGB_R、RGB_G 和 RGB_B 两端形成压降，RGB 灯点亮。

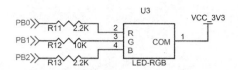

图 3.11　RGB 灯与 ZI-ARMEmbed 的连接

按键与 ZI-ARMEmbed 的连接同 2.5.2.1 节。

3.3.2.2　软件设计

软件设计流程如图 3.12 所示。

（1）在主函数中完成 RGB 灯和按键的初始化。

（2）判断按键是否按下，根据按下的按键点亮不同颜色的 RGB 灯，并在 MobaXterm 串口终端 FinSH 控制台输出提示信息。按下 KEY1，FinSH 控制台输出"Press the button number: 1"，同时点亮 RGB 红灯；按下 KEY2，FinSH 控制台输出 "Press the button number: 2"，同时点亮 RGB 绿灯；按下 KEY3，FinSH 控制台输出 "Press the button number: 3"，同时点亮 RGB 蓝灯。

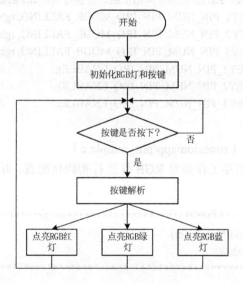

图 3.12　软件设计流程

3.3.2.3　功能设计与核心代码设计

要控制 PIN 设备，需要使用 RT-Thread 提供的 PIN 设备管理接口对 PIN 设备进行初始化。

1）主函数（zonesion/app/main.c）

主函数的主要工作是初始化相关设备。代码如下：

```
/**************************************************************************
* 名称：main()
* 功能：初始化相关设备
**************************************************************************/
#include "pin_sample.h"
#include "pin_sensor.h"
int main(void)
{
```

```
    key_int_init();                                              //按键及其中断的初始化
    rgb_pin_init();                                              //RGB 灯的初始化
    return 0;
}
```

2）按键初始化函数（zonesion/app/pin_sample.c）

按键初始化函数的主要工作是对按键进行初始化配置，并为其设置中断，代码如下：

```
/*****************************************************************************
* 名称：key_int_init()
* 功能：按键初始化
*****************************************************************************/
void key_int_init(void)
{
    rt_pin_mode(KEY1_PIN_NUM, PIN_MODE_INPUT_PULLUP);        //将 KEY1 设置为上拉输入
    rt_pin_mode(KEY2_PIN_NUM, PIN_MODE_INPUT_PULLUP);        //将 KEY2 设置为上拉输入
    rt_pin_mode(KEY3_PIN_NUM, PIN_MODE_INPUT_PULLUP);        //将 KEY3 设置为上拉输入
    //设置 KEY1、KEY2、KEY3 引脚下降沿中断，并执行 rgb_ctrl 函数
    rt_pin_attach_irq(KEY1_PIN_NUM, PIN_IRQ_MODE_FALLING, rgb_ctrl, (void*)0x01);
    rt_pin_attach_irq(KEY2_PIN_NUM, PIN_IRQ_MODE_FALLING, rgb_ctrl, (void*)0x02);
    rt_pin_attach_irq(KEY3_PIN_NUM, PIN_IRQ_MODE_FALLING, rgb_ctrl, (void*)0x03);
    rt_pin_irq_enable(KEY1_PIN_NUM, PIN_IRQ_ENABLE);          //使能 KEY1 中断
    rt_pin_irq_enable(KEY2_PIN_NUM, PIN_IRQ_ENABLE);          //使能 KEY2 中断
    rt_pin_irq_enable(KEY3_PIN_NUM, PIN_IRQ_ENABLE);          //使能 KEY3 中断
}
```

3）RGB 初始化函数（zonesion/app/pin_sample.c）

RGB 初始化函数的主要工作是对 RGB 灯进行初始化配置，并将其引脚默认设置为高电平，代码如下：

```
/*****************************************************************************
* 名称：rgb_pin_init (void *parameter)
* 功能：RGB 灯初始化函数
*****************************************************************************/
void rgb_pin_init(void)
{
    rt_pin_mode(RED_PIN_NUM, PIN_MODE_OUTPUT);              //将 RGB 红灯的引脚设置为输出模式
    rt_pin_mode(GREEN_PIN_NUM, PIN_MODE_OUTPUT);            //将 RGB 绿灯的引脚设置为输出模式
    rt_pin_mode(BLUE_PIN_NUM, PIN_MODE_OUTPUT);             //将 RGB 蓝灯的引脚设置为输出模式
    rt_pin_write(RED_PIN_NUM, PIN_HIGH);                    //将 RGB 红灯的引脚设置为高电平
    rt_pin_write(GREEN_PIN_NUM, PIN_HIGH);                  //将 RGB 绿灯的引脚设置为高电平
    rt_pin_write(BLUE_PIN_NUM, PIN_HIGH);                   //将 RGB 蓝灯的引脚设置为高电平
}
```

4）自定义 FinSH 控制台命令函数（zonesion/app/finsh_cmd/finsh_ex.c）

自定义 FinSH 控制台命令函数的主要工作是实现 PIN 设备管理的命令。代码如下：

```
/*************************************************************************
 * 名称：threadManage(int argc, char **argv)
 * 功能：自定义 FinSH 控制台命令
 * 参数：argc 表示参数个数；argv 表示指向参数字符串指针的数组指针
 *************************************************************************/
#include <rtthread.h>
#include <string.h>
#include "pin_sensor.h"
int pinSensor(int argc, char **argv)
{
    if(!rt_strcmp(argv[1], "get"))
    {
        t_pin_sensor_status sensor_status;
        pin_sensor_read(&sensor_status);
        rt_kprintf("infrared:\t%u\n", sensor_status.infrared);
        rt_kprintf("vibration:\t%u\n", sensor_status.vibration);
        rt_kprintf("hall:\t\t%u\n", sensor_status.hall);
        rt_kprintf("flame:\t\t%u\n", sensor_status.flame);
        rt_kprintf("grating:\t%u\n", sensor_status.grating);
        rt_kprintf("fan:\t\t%u\n", sensor_status.fan);
        rt_kprintf("lock:\t\t%u\n", sensor_status.lock);
    } else if(!rt_strcmp(argv[1], "fan")) {
        if(!rt_strcmp(argv[2], "on")) {
            pin_sensor_ctrl(FAN_PIN_NUM, 1);
            rt_kprintf("fan on\n");
        } else if(!rt_strcmp(argv[2], "off")) {
            pin_sensor_ctrl(FAN_PIN_NUM, 0);
            rt_kprintf("fan off\n");
        }
    } else if(!rt_strcmp(argv[1], "lock")) {
        if(!rt_strcmp(argv[2], "open")) {
            pin_sensor_ctrl(LOCK_PIN_NUM, 1);
            rt_kprintf("lock open\n");
        } else if(!rt_strcmp(argv[2], "close")) {
            pin_sensor_ctrl(LOCK_PIN_NUM, 0);
            rt_kprintf("lock close\n");
        }
    }
    return 0;
}
MSH_CMD_EXPORT(pinSensor, pinSensor Read and Control);
```

3.3.3　开发步骤与验证

3.3.3.1　硬件部署

同 1.2.3.1 节。

3.3.3.2　工程调试

（1）将本项目的工程（16-PIN）文件夹复制到 RT-ThreadStudio\workspace 目录下。

（2）其余同 2.1.3.2 节。

3.3.3.4　验证效果

（1）关闭 RT-Thread Studio，拔掉仿真器，按下 ZI-ARMEmbed 上的电源按键重新上电。

（2）按下 KEY1，MobaXterm 串口终端 FinSH 控制台输出 "Press the button number: 1"，同时点亮 RGB 红灯。

```
 \ | /
- RT -     Thread Operating System
 / | \     4.1.0 build Oct 21 2022 11:41:40
 2006 - 2022 Copyright by RT-Thread team
Press the button number: 1
```

（3）按下 KEY2，MobaXterm 串口终端 FinSH 控制台输出 "Press the button number: 2"，同时点亮 RGB 绿灯。

```
 \ | /
- RT -     Thread Operating System
 / | \     4.1.0 build Oct 21 2022 11:41:40
 2006 - 2022 Copyright by RT-Thread team
Press the button number: 1
Press the button number:2
```

（4）按下 KEY3，MobaXterm 串口终端 FinSH 控制台输出 "Press the button number: 3"，同时点亮 RGB 蓝灯。

```
 \ | /
- RT -     Thread Operating System
 / | \     4.1.0 build Oct 21 2022 11:41:40
 2006 - 2022 Copyright by RT-Thread team
Press the button number: 1
Press the button number: 2
Press the button number: 3
```

3.3.4　小结

本节主要介绍 PIN 设备模型框架的基本概念和管理方式。通过本节的学习，读者可掌握 RT-Thread PIN 设备管理接口的应用。

3.4 ADC 设备驱动应用开发

模数转换器（Analog-to-Digital Converter，ADC）用于将模拟信号转换为数字信号，这种转换通常应用在需要处理数字信号的系统中，如微控制器、DSP 等。在测量和采集模拟信号（如温度、光强、声音）等场景，ADC 得到了广泛的应用。

本节的要求如下：

- ➲ 了解 ADC 设备的基本概念。
- ➲ 掌握 ADC 设备的管理方式。
- ➲ 掌握 RT-Thread ADC 设备管理接口的应用。

3.4.1　原理分析

3.4.1.1　ADC 设备简介

ADC（Analog-to-Digital Converter）指模数转换器，是一种将模拟信号转换为数字信号的器件。

真实世界的模拟信号，如温度、压力、声音或者图像等，往往需要转换成更容易存储、处理和传输的数字信号。ADC 可实现这个转换功能，在很多产品中都可以找到 ADC 的身影。

模数转换过程如图 3.13 所示，一般要经过采样、保持、量化、编码这几个步骤。在实际电路中，有些步骤是合并进行的，如采样和保持、量化和编码在转换过程中是同时实现的。

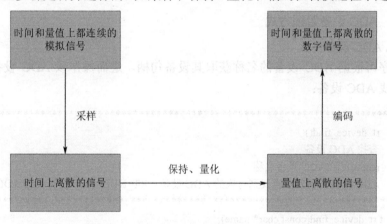

图 3.13　模数转换过程

采样是指将时间上连续变化的模拟信号转换为时间上离散的模拟信号。将采样得到的模拟信号转换为数字信号需要一定时间，为了给后续的量化、编码过程提供一个稳定的值，需要将采样得到的模拟信号保持一段时间。

将数值连续的模拟信号转换为数字信号的过程称为量化。数字信号在数值上是离散的，采样、保持电路的输出电压需要按照某种近似方式转换成相应的离散电平。任何数字量只能是某个最小数量单位的整数倍。量化后的数值还需要进行编码，编码结果就是 ADC 输出的数字信号。

与 ADC 有关的概念与原理如下：

（1）模拟信号与数字信号。模拟信号是连续的信号，其数值在一定范围内可以取任何值，如声音的波形、温度传感器输出的电压等。数字信号是离散的信号，其数值通常是以二进制数字形式表示的，ADC 将模拟信号转换为数字信号，可方便数字系统进行处理。

（2）采样和量化。ADC 通过一定的时间间隔内对模拟信号进行采样，可得到一系列离散的样本点。采样后，模拟信号的量值被量化为离散的数字值。量化决定了 ADC 的分辨率。

（3）ADC 的分辨率。分辨率用来表示 ADC 能够区分的模拟信号的最小变化量，通常以比特（bit）表示。例如，8 位 ADC 可以表示 256 个不同的离散值。分辨率通常是以二进制（或十进制）数的位数来表示的，一般有 8 位、10 位、12 位、16 位等，用于表明 ADC 对输入信号的分辨能力，位数越多，分辨率越高，恢复的模拟信号会更精确。

（4）ADC 的采样速率。采样速率是指 ADC 在 1 s 内进行采样的次数，高采样速率的 ADC 通常用于捕捉高频信号，但会导致更多的数据需要处理。

（5）ADC 的类型。逐次逼近型 ADC 是一种常见的 ADC 类型，它通过逐渐逼近输入信号的值来完成转换。逐次逼近型 ADC 会按顺序处理多个位，逐渐逼近模拟信号的值。并行型 ADC 可以同时处理多个位，可提高转换速率。

（6）ADC 的应用领域。

- ◒ 传感器接口：ADC 可将传感器的模拟输出转换为数字形式，如温度传感器、光敏传感器等。
- ◒ 音频处理：在音频处理中，ADC 可将模拟的声音波形转换为数字音频信号。
- ◒ 通信系统：在通信系统中，ADC 可将模拟信号（如语音）转换为数字信号，以便通过数字通信系统进行传输。

3.4.1.2　ADC 设备管理方式

1）查找 ADC 设备

应用程序可根据 ADC 设备的名称获取其设备句柄，进而操作该 ADC 设备。通过下面的函数可查找 ADC 设备：

```
/******************************************************************************
 * 名称：rt_device_find()
 * 功能：查找 ADC 设备
 * 参数：name 表示 ADC 设备名称
 * 返回：ADC 设备句柄表示查找到对应的 ADC 设备；RT_NULL 表示没有找到 ADC 设备
 ******************************************************************************/
rt_device_t rt_device_find(const char* name);
```

2）使能 ADC 设备通道

在读取 ADC 设备通道的输出数据前需要先使能该 ADC 设备通道。通过下面的函数可使能 ADC 设备通道：

```
/******************************************************************************
 * 名称：rt_adc_enable()
 * 功能：使能 ADC 设备通道
 * 参数：dev 表示 ADC 设备句柄；channel 表示 ADC 设备通道；
 * 返回：RT_EOK 表示成功；-RT_ENOSYS 表示 ADC 设备的操作方法为空；其他错误码表示失败
 ******************************************************************************/
rt_err_t rt_adc_enable(rt_adc_device_t dev, rt_uint32_t channel);
```

3）读取 ADC 设备通道的采样值

通过下面的函数可读取 ADC 设备通道的采样值（输出数据）：

```
/******************************************************************************
 * 名称：rt_adc_read()
 * 功能：读取 ADC 设备通道的采样值
```

```
* 参数：dev 表示 ADC 设备句柄；channel 表示 ADC 设备通道
* 返回：读取的数据
*******************************************************************************/
rt_uint32_t rt_adc_read(rt_adc_device_t dev, rt_uint32_t channel);
```

4）关闭 ADC 设备通道

通过下面的函数可关闭 ADC 设备通道：

```
/*******************************************************************************
* 名称：rt_adc_disable()
* 功能：关闭 ADC 设备通道
* 参数：dev 表示 ADC 设备句柄
* 返回：RT_EOK 表示成功；-RT_ENOSYS 表示 ADC 设备操作方法为空；其他错误码表示失败
*******************************************************************************/
rt_err_t rt_adc_disable(rt_adc_device_t dev, rt_uint32_t channel);
```

3.4.2　开发设计与实践

3.4.2.1　硬件设计

本节的硬件设计同 3.2.2.1 节。本节通过 FinSH 控制台实现了对 ADC 设备的控制，有助于读者掌握 ADC 设备管理接口的应用。

3.4.2.2　软件设计

软件设计流程如图 3.14 所示。

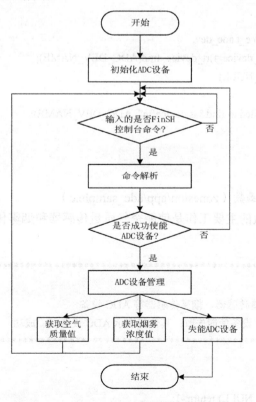

图 3.14　软件设计流程

（1）初始化 ADC 设备，并且在主函数中查找 ADC 设备。

（2）在 MobaXterm 串口终端 FinSH 控制台中输入"adcDrv open"命令，使能 ADC 设备。当在 FinSH 控制台中输入"adcDrv get air"命令时，可获取空气质量值；当在 FinSH 控制台中输入"adcDrv get gas"命令时，可获取烟雾浓度值。当在 FinSH 控制台中输入"adcDrv close"命令时，可失能 ADC 设备。

3.4.2.3　功能设计与核心代码设计

要控制 ADC 设备，首先需要使用 RT-Thread 提供的 ADC 设备管理接口进行 ADC 设备的初始化，然后使用 FinSH 控制台命令控制 ADC 设备。需要用户实现对应的 FinSH 控制台命令，并将已实现的命令添加到 FinSH 控制台命令列表中，这样在启动 FinSH 控制台后就可以使用自定义的命令了。

1）主函数（zonesion/app/main.c）

主函数的主要工作是查找 ADC 设备，完成之后退出主函数。代码如下：

```
/*****************************************************************************
* 名称：main()
* 功能：查找 ADC 设备
*****************************************************************************/
#include <rtthread.h>
#include "drivers/adc.h"
#include "adc_sample.h"
int main(void)
{
    extern rt_adc_device_t adc_dev;
    adc_dev = (rt_adc_device_t)rt_device_find(ADC_DEV_NAME);
    if(adc_dev == RT_NULL)
    {
        rt_kprintf("failed to find %s device!\n", ADC_DEV_NAME);
    }
    return 0;
}
```

2）ADC 设备使能函数（zonesion/app/adc_sample.c）

ADC 设备使能函数的主要工作是使能空气质量传感器和烟雾传感器等 ADC 设备。代码如下：

```
/*****************************************************************************
* 名称：adc_open()
* 功能：使能空气质量传感器、烟雾传感器等 ADC 设备
* 返回：-1 表示 ADC 设备使能失败；其他值表示 ADC 设备使能成功
*****************************************************************************/
int adc_open(void)
{
    if(adc_dev == RT_NULL) return -1;
    if(rt_adc_enable(adc_dev, AIR_ADC_CHANNEL)==RT_EOK &&
```

```
                                        rt_adc_enable(adc_dev, GAS_ADC_CHANNEL)==RT_EOK) {
        return 0;
    }
    return -1;
}
```

3）获取 ADC 设备返回值函数（zonesion/app/adc_sample.c）

获取 ADC 设备返回值函数的主要工作是获取空气质量值和烟雾浓度值。代码如下：

```
/**************************************************************************
 * 名称：adc_getVal(uint8_t channel)
 * 功能：获取 ADC 设备返回值
 * 参数：channel 表示要获取的 ADC 设备通道
 * 返回：-1 表示 ADC 设备未打开；其他值表示 ADC 设备的返回值
 **************************************************************************/
int adc_getVal(uint8_t channel)
{
    if(adc_dev != RT_NULL)
    {
        int value = 0;
        if(channel == AIR_ADC_CHANNEL || channel == GAS_ADC_CHANNEL) {
            value = rt_adc_read(adc_dev, channel);
            rt_kprintf("ADC value: %d\n", value);
            return value;
        }
    }
    rt_kprintf("ADC drive not open!\n");
    return -1;
}
```

4）失能 ADC 设备函数（zonesion/app/adc_sample.c）

失能 ADC 设备函数的主要工作是关闭空气质量传感器、烟雾传感器等 ADC 设备。代码如下：

```
/**************************************************************************
 * 名称：adc_close()
 * 功能：关闭 ADC 设备
 * 返回：-1 表示关闭失败，0 表示关闭成功
 **************************************************************************/
int adc_close(void)
{
    if(adc_dev == RT_NULL)
    {
        rt_kprintf("ADC drive not on!\n");
        return -1;
    }
    rt_err_t result = RT_ERROR;
```

```
       if(rt_adc_disable(adc_dev, AIR_ADC_CHANNEL)==RT_EOK &&
                              rt_adc_disable(adc_dev, GAS_ADC_CHANNEL)==RT_EOK) {
           result = RT_EOK;
       }
       if(result == RT_EOK)
           adc_dev = RT_NULL;
       return result;
   }
```

5）自定义 FinSH 控制台命令函数（zonesion/app/finsh_cmd/finsh_ex.c）

自定义 FinSH 控制台命令函数的主要工作是实现 ADC 设备管理的命令，包括使能 ADC 设备、获取 ADC 设备返回值、失能 ADC 设备等，可以在 MobaXterm 串口终端 FinSH 控制台中使用相关命令对 ADC 设备进行操作。代码如下：

```
/*****************************************************************************
 * 名称：adcDrv (int argc, char **argv)
 * 功能：自定义 FinSH 控制台命令
 * 参数：argc 表示参数个数；argv 表示指向参数字符串指针的数组指针
 *****************************************************************************/
#include <rtthread.h>
#include "adc_sample.h"

int adcDrv(int argc, char **argv)
{
    rt_err_t result = RT_EOK;
    if(!rt_strcmp(argv[1], "open"))
    {
        result = adc_open();                                    //使能 ADC 设备
        if(result == RT_EOK)
            rt_kprintf("adc enable success!\n");
        else
            rt_kprintf("adc enable failed, errCode:%d\n", result);
    }
    else if(!rt_strcmp(argv[1], "get"))
    {
        if(!rt_strcmp(argv[2], "air")) {
            adc_getVal(AIR_ADC_CHANNEL);
        } else if(!rt_strcmp(argv[2], "gas")) {
            adc_getVal(GAS_ADC_CHANNEL);
        }
    }
    else if(!rt_strcmp(argv[1], "close"))
    {
        result = adc_close();                                  //失能 ADC 设备
        if(result == RT_EOK)
            rt_kprintf("adc disable success!\n");
        else
            rt_kprintf("adc disable failed, errCode:%d\n", result);
```

```
    }
    else
        rt_kprintf("Please input 'adcDrv <open|get|close>'\n");
    return 0;
}
MSH_CMD_EXPORT(adcDrv, adc driver sample);
```

3.4.3　开发步骤与验证

3.4.3.1　硬件部署

同 1.2.3.1 节。

3.4.3.2　工程调试

（1）将本项目的工程（17-ADC）文件夹复制到 RT-ThreadStudio\workspace 目录下。

（2）其余同 2.1.3.2 节。

3.4.3.3　验证效果

（1）关闭 RT-Thread Studio，拔掉仿真器，按下 ZI-ARMEmbed 上的电源按键重新上电。

（2）在 MobaXterm 串口终端 FinSH 控制台中输入"adcDrv open"命令，解析自定义 FinSH 控制台命令后可使能 ADC 设备，FinSH 控制台输出"adc enable success!"。

```
 \ | /
- RT -     Thread Operating System
 / | \     4.1.0 build Oct 21 2022 17:46:18
 2006 - 2022 Copyright by RT-Thread team
msh >adcDrv open
adc enable success!
```

（3）在 MobaXterm 串口终端 FinSH 控制台中输入"adcDrv get air"命令，解析自定义 FinSH 控制台命令后可获取空气质量值，FinSH 控制台输出"ADC value: 12"。

```
 \ | /
- RT -     Thread Operating System
 / | \     4.1.0 build Oct 21 2022 17:46:18
 2006 - 2022 Copyright by RT-Thread team
msh >adcDrv open
adc enable success!
msh >adcDrv get air
ADC value: 12
```

（4）在 MobaXterm 串口终端 FinSH 控制台中输入"adcDrv get gas"命令，解析自定义 FinSH 控制台命令后可获取烟雾浓度值，FinSH 控制台输出"ADC value: 132"。

```
 \ | /
- RT -     Thread Operating System
 / | \     4.1.0 build Oct 21 2022 17:46:18
 2006 - 2022 Copyright by RT-Thread team
msh >adcDrv open
```

```
adc enable success!
msh >adcDrv get air
ADC value: 12
msh >adcDrv get gas
ADC value: 132
```

（5）在 MobaXterm 串口终端 FinSH 控制台中输入"adcDrv close"命令，解析自定义 FinSH 控制台命令后可关闭 ADC 设备，FinSH 控制台输出"adc disable success!"。

```
  \ | /
- RT -     Thread Operating System
 / | \     4.1.0 build Oct 21 2022 17:46:18
 2006 - 2022 Copyright by RT-Thread team
msh >adcDrv open
adc enable success!
msh >adcDrv get air
ADC value: 12
msh >adcDrv get gas
ADC value: 132
msh >adcDrv close
adc disable success!
```

3.4.4 小结

本节主要介绍 ADC 设备的基本概念和管理方式。通过本节的学习，读者可掌握 RT-Thread ADC 设备管理接口的应用。

3.5 HWTIMER 设备驱动应用开发

硬件定时器（Hardware Timer，HWTIMER）设备是一种常用的嵌入式系统硬件设备，用于生成精确的延时、定时触发任务或执行周期性的操作。HWTIMER 设备通常由微控制器或微处理器提供，其主要功能包括计时、计数和触发中断。

本节的要求如下：
- 了解 HWTIMER 设备的基本概念。
- 掌握 HWTIMER 设备的管理方式。
- 掌握 RT-Thread HWTIMER 设备管理接口的应用。

3.5.1 原理分析

3.5.1.1 HWTIMER 设备简介

HWTIMER 设备一般有两种工作模式，即定时器模式和计数器模式。不管 HWTIMER 设备工作在哪种模式，实质都是通过内部定时器模块来对脉冲信号进行计数的。

与 HWTIMER 设备有关的概念如下：

（1）计数器模式：对输入引脚的外部脉冲信号进行计数。

（2）定时器模式：对内部脉冲信号进行计数。HWTIMER 设备常用于定时检测、定时响应、定时控制。

（3）定时器：定时器可以递增计数或递减计数。16 位定时器的最大计数值为 65535，32 位定时器的最大值为 4294967295。

（4）计数频率：在定时器模式下，由于系统时钟频率是固定值，所以可以根据定时器的计数值计算出定时时间，定时时间=计数值/计数频率。例如，计数频率为 1 MHz，定时器计数一次的时间则为 1/1000000，也就是每经过 1 μs 定时器就加 1（或减 1），此时 16 位定时器的最大定时能力为 65535 μs，即 65.535 ms。

3.5.1.2　HWTIMER 设备特点和应用

1）特点

- 精确性：HWTIMER 通常非常精确，因为它们是硬件设备，不受操作系统或其他软件的影响。
- 高分辨率：HWTIMER 设备通常具有高分辨率，可以提供微秒级别的时间精度。
- 计数功能：HWTIMER 可以执行计数操作，通常用于测量时间间隔或执行定时操作。
- 中断触发：HWTIMER 通常可以在特定时间间隔或特定计数值时触发中断，从而执行与时间相关的任务。

2）应用

- 实时操作系统（RTOS）：HWTIMER 设备通常在 RTOS 中用于任务调度、时间片管理和触发任务的执行。
- 通信协议：在通信系统中，HWTIMER 设备可用于生成精确的时间间隔，以进行数据传输、同步和时隙分配。
- 嵌入式控制系统：在嵌入式系统中，HWTIMER 设备用于实现精确的控制循环，如电机控制、传感器数据采集和执行特定操作。
- 通信接口：在通信接口中，HWTIMER 设备可用于生成特定的时序信号，如 UART、SPI 和 I2C 通信中的波特率时钟。
- 精密测量：HWTIMER 设备可用于精密时间测量，如测量信号的脉冲宽度、计算速度或周期性事件的间隔。

3.5.1.3　HWTIMER 设备的管理方式

1）查找 HWTIMER 设备

应用程序可根据 HWTIMER 设备名称获取该设备句柄，进而操作 HWTIMER 设备。通过下面的函数可以查找 HWTIMER 设备：

```
/************************************************************************
 * 名称：rt_device_find()
 * 功能：查找 HWTIMER 设备
 * 参数：name 表示 HWTIMER 设备名称
 * 返回：HWTIMER 设备句柄表示查找到对应的设备；RT_NULL 表示没有找到对应的设备
 ************************************************************************/
```

```
rt_device_t rt_device_find(const char* name);
```

2）打开 HWTIMER 设备

通过 HWTIMER 设备句柄，应用程序可以打开 HWTIMER 设备。在打开 HWTIMER 设备时，RT-Thread 会检测 HWTIMER 设备是否已经初始化，如果没有初始化则会调用默认的初始化接口初始化 HWTIMER 设备。通过下面的函数可打开 HWTIMER 设备：

```
/********************************************************************************
 * 名称：rt_device_open()
 * 功能：打开 HWTIMER 设备
 * 参数：dev 表示 HWTIMER 设备句柄；oflags 表示 HWTIMER 设备的打开模式，一般以读写方式打
开，即该参数的取值为 RT_DEVICE_OFLAG_RDWR
 * 返回：RT_EOK 表示 HWTIMER 设备打开成功；其他错误码表示 HWTIMER 设备打开失败
 ********************************************************************************/
rt_err_t rt_device_open(rt_device_t dev, rt_uint16_t oflags);
```

3）设置超时回调函数

通过下面的函数可设置超时回调函数，当 HWTIMER 设备超时将会调用回调函数：

```
/********************************************************************************
 * 名称：rt_device_set_rx_indicate()
 * 功能：设置超时回调函数
 * 参数：dev 表示 HWTIMER 设备句柄；rx_ind 表示超时回调函数，由调用者提供
 * 返回：RT_EOK 表示设置成功
 ********************************************************************************/
rt_err_t rt_device_set_rx_indicate(rt_device_t dev, rt_err_t (*rx_ind)(rt_device_t dev,rt_size_t size));
```

4）控制 HWTIMER 设备

应用程序可通过命令控制字对 HWTIMER 设备进行配置，通过下面的函数可控制 HWTIMER 设备：

```
/********************************************************************************
 * 名称：rt_device_control()
 * 功能：控制 HWTIMER 设备
 * 参数：dev 表示 HWTIMER 设备句柄；cmd 表示命令控制字；arg 表示控制的参数
 * 返回：RT_EOK 表示函数执行成功；-RT_ENOSYS 表示执行失败（dev 为空）；其他错误码表示
执行失败
 ********************************************************************************/
rt_err_t rt_device_control(rt_device_t dev, rt_uint8_t cmd, void* arg);
```

5）设置 HWTIMER 设备超时值

通过下面的函数可设置 HWTIMER 设备超时值（超时时间）：

```
/********************************************************************************
 * 名称：rt_device_write()
 * 功能：设置 HWTIMER 设备超时值
 * 参数：dev 表示 HWTIMER 设备句柄；pos 表示写入数据偏移量，未使用，可取 0；buffer 表示指
向 HWTIMER 设备超时时间结构体的指针；size 表示超时时间结构体的大小
 * 返回：写入数据的实际大小表示设置成功；0 表示设置失败
 ********************************************************************************/
rt_size_t rt_device_write(rt_device_t dev, rt_off_t pos, const void* buffer, rt_size_t size);
```

6）获取 HWTIMER 设备当前值

通过下面的函数可获取 HWTIMER 设备的当前值（当前时间）：

```
/*****************************************************************************
 * 名称：rt_device_read()
 * 功能：获取 HWTIMER 设备当前值
 * 参数：dev 表示 HWTIMER 设备句柄；pos 表示写入数据偏移量，未使用，可取 0；buffer 表示输
出参数，指向 HWTIMER 设备超时时间结构体的指针；size 表示超时时间结构体的大小
 * 返回：超时时间结构体的大小表示成功；0 表示获取失败
 *****************************************************************************/
rt_size_t rt_device_read(rt_device_t dev, rt_off_t pos, void* buffer, rt_size_t size);
```

使用示例如下所示：

```
rt_hwtimerval_t t;
//读取 HWTIMER 设备当前值
rt_device_read(hw_dev, 0, &t, sizeof(t));
rt_kprintf("Read: Sec = %d, Usec = %d\n", t.sec, t.usec);
```

3.5.2　开发设计与实践

3.5.2.1　硬件设计

本节的硬件设计同 3.2.2.1 节。本节通过 MobaXterm 串口终端 FinSH 控制台控制 HWTIMER 设备，帮助读者掌握 HWTIMER 设备管理接口的应用。

3.5.2.2　软件设计

软件设计流程如图 3.15 所示。

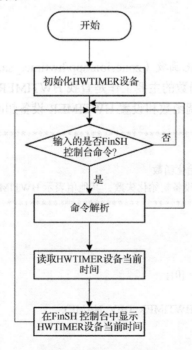

图 3.15　软件设计流程

（1）在主函数中初始化 HWTIMER 设备。

（2）编写 HWTIMER 设备超时回调函数，在 FinSH 控制台显示 HWTIMER 设备名称和当前时间。

（3）在 MobaXterm 串口终端 FinSH 控制台中输入"hwtimerDrv read"命令，可读取 HWTIMER 设备当前时间，并将返回信息显示在 FinSH 控制台中。

3.5.2.1 功能设计与核心代码设计

要控制 HWTIMER 设备，首先需要使用 RT-Thread 提供的 HWTIMER 设备管理接口来初始化 HWTIMER 设备，然后编写 HWTIMER 设备超时回调函数。本节使用 FinSH 控制台对 HWTIMER 设备进行控制，需要自定义 FinSH 控制台命令，并将已实现命令添加到 FinSH 控制台命令列表中，这样在启动 MobaXterm 串口终端 FinSH 控制台后就可以使用自定义 FinSH 控制台命令了。

1）主函数（zonesion/app/main.c）

主函数的主要工作是实现 HWTIMER 设备初始化，初始化完成之后退出主函数。代码如下：

```
/*************************************************************************
* 名称：main()
* 功能：HWTIMER 设备初始化
*************************************************************************/
#include "hwtimer_sample.h"
#include "drv_led.h"
int main(void)
{
    hwtimer_init();
    return 0;
}
```

2）HWTIMER 设备初始化函数（zonesion/app/hwtimer_sample.c）

HWTIMER 设备初始化函数的主要工作是查找 HWTIMER 设备、打开 HWTIMER 设备、设置 HWTIMER 设备超时回调函数和设置 HWTIMER 设备超时时间等。代码如下：

```
/*************************************************************************
* 名称：hwtimer_init()
* 功能：HWTIMER 设备初始化函数
* 返回：-1 表示 HWTIMER 设备初始化失败；其他值表示 HWTIMER 设备初始化成功
*************************************************************************/
int hwtimer_init(void)
{
    rt_err_t result= RT_EOK;
    rt_hwtimerval_t timeOut = {0};
    rt_hwtimer_mode_t mode;
    hw_dev = rt_device_find(HWTIMER_DEV_NAME);              //查找 HWTIMER 设备
    if(hw_dev == RT_NULL)
    {
        rt_kprintf("failed to find %s device!\n", HWTIMER_DEV_NAME);
```

```
            return -1;
        }
        result = rt_device_open(hw_dev, RT_DEVICE_OFLAG_RDWR);      //打开 HWTIMER 设备
        if(result != RT_EOK)
        {
            rt_kprintf("open %s device failed!\n", HWTIMER_DEV_NAME);
            return -1;
        }
        rt_device_set_rx_indicate(hw_dev, hwtimer_cb);              //设置 HWTIMER 设备超时回调函数
        mode = HWTIMER_MODE_PERIOD;
        result = rt_device_control(hw_dev, HWTIMER_CTRL_MODE_SET, &mode);//设置为周期性定时器模式
        if (result != RT_EOK)
        {
            rt_kprintf("set mode failed! errCode: %d\n", result);
            return result;
        }
        timeOut.sec = 5;                                           //5 s
        timeOut.usec = 0;                                          //0 μs
        //设置 HWTIMER 设备超时时间
        if (rt_device_write(hw_dev, 0, &timeOut, sizeof(timeOut)) != sizeof(timeOut))
        {
            rt_kprintf("set timeout value failed!\n");
            return RT_ERROR;
        }
        return result;
}
```

3）HWTIMER 设备超时回调函数（zonesion/app/hwtimer_sample.c）

HWTIMER 设备超时回调函数的主要工作是在 HWTIMER 设备超时后给出相应的提示信息。代码如下：

```
/***************************************************************************
* 名称：hwtimer_cb()
* 功能：HWTIMER 设备超时回调函数
* 参数：dev 表示设备句柄；size 表示 HWTIMER 设备超时时间结构体的大小
***************************************************************************/
static rt_err_t hwtimer_cb(rt_device_t dev, rt_size_t size)
{
    rt_kprintf("device name: %s\n", dev->parent.name);
    rt_kprintf("tick is:%d\n", rt_tick_get());
    return RT_EOK;
}
```

4）自定义 FinSH 控制台命令函数模块（zonesion/app/finsh_cmd/finsh_ex.c）

自定义 FinSH 控制台命令函数的主要工作是实现 HWTIMER 设备的管理命令，可在 FinSH 控制台中使用这些命令来读取 HWTIMER 设备当前时间。代码如下：

```
/***************************************************************************
* 名称：hwtimerDrv (int argc, char **argv)
```

```
 * 功能：自定义 FinSH 控制台命令
 * 参数：argc 表示参数个数；argv 表示指向参数字符串指针的数组指针
 ***********************************************************************************/
#include <rtthread.h>
#include "hwtimer_sample.h"
int hwtimerDrv(int argc, char **argv)
{
    rt_hwtimerval_t timeOut = {0};
    if(!rt_strcmp(argv[1], "read"))
    {
        rt_device_read(hw_dev, 0, &timeOut, sizeof(timeOut));     //读取 HWTIMER 设备当前时间
        rt_kprintf("read: sec = %d, usec = %d\n", timeOut.sec, timeOut.usec);
    }
    else
        rt_kprintf("Please input 'hwtimerDrv <read>'\n");
    return 0;
}
MSH_CMD_EXPORT(hwtimerDrv, hwtimer driver sample);
```

3.5.3　开发步骤与验证

3.5.3.1　硬件部署

同 1.2.3.1 节。

3.5.3.2　项目部署

（1）将本项目的工程（18-HWTIMER）文件夹复制到 RT-ThreadStudio\workspace 目录下。
（2）其余同 2.1.3.2 节。

3.5.3.3　验证效果

（1）关闭 RT-Thread Studio，拔掉仿真器，按下 ZI-ARMEmbed 上的电源按键重新上电。
（2）在 MobaXterm 串口终端 FinSH 控制台中输出 HWTIMER 设备名称"device name: timer3"和 HWTIMER 设备触发的滴答计数"tick is:5001"。

```
 \ | /
- RT -     Thread Operating System
 / | \       4.1.0 build Oct 21 2022 17:59:18
 2006 - 2022 Copyright by RT-Thread team
device name: timer3
tick is:5001
```

（3）在 MobaXterm 串口终端 FinSH 控制台中输入"hwtimerDrv read"命令，解析自定义 FinSH 控制台命令后可读取 HWTIMER 设备的计时时间，并在 FinSH 控制台中输出"read: sec = 8, usec = 878691"。

```
 \ | /
- RT -     Thread Operating System
 / | \       4.1.0 build Oct 21 2022 17:59:18
```

device name: timer3
tick is:5001
msh >**hwtimerDrv read**
read: sec = 8, usec = 878691

3.5.4　小结

本节主要介绍 HWTIMER 设备的基本概念、特点、应用和管理方式。通过本节的学习，读者可掌握 RT-Thread HWTIMER 设备管理接口的应用。

3.6 I2C 设备驱动应用开发

I2C（Inter-Integrated Circuit）总线是一种串行通信协议，用于连接微控制器和各种外部设备，如传感器、显示屏、存储器芯片等。I2C 总线是由 Philips（现在的 NXP）于 1982 年提出的，是一种半双工、双向二线制同步串行总线，支持多主机和多从机的连接。I2C 总线的应用领域包括：

- ➲ 传感器和芯片：I2C 总线常用于连接各种传感器和芯片，如温度传感器、加速度计、EEPROM 等。
- ➲ 显示器：I2C 总线可用于连接液晶显示器（LCD）等显示设备。
- ➲ 通信设备：在通信设备中，I2C 总线可用于连接蓝牙模块、Wi-Fi 模块等。

I2C 总线的优点是只需要两根信号线即可传输数据，支持多主机和多从机，适用于短距离通信等。在使用 I2C 设备时，配置正确的时钟频率、地址、开始信号和停止信号是非常重要的。

本节的要求如下：

- ➲ 了解 I2C 设备的基本概念。
- ➲ 掌握 I2C 设备的管理方式。
- ➲ 掌握 RT-Thread I2C 设备管理接口的应用。

3.6.1　原理分析

3.6.1.1　I2C 设备简介

I2C 总线在传输数据时只需要两根信号线，一根是双向数据线 SDA（Serial Data），另一根是双向时钟线 SCL（Serial Clock）。I2C 总线和 SPI 总线一样，采用的都是主从工作方式，不同于 SPI 总线的一主多从结构，I2C 总线允许同时存在多个主机，每个连接到总线上的设备都有唯一的地址，主机启动数据传输并产生时钟信号，从机被主机寻址，同一时刻只允许有一个主机。I2C 总线的连接如图 3.16 所示。

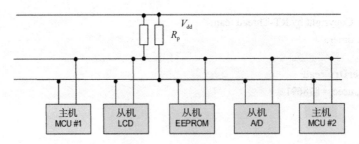

图 3.16 I2C 总线的连接

I2C 总线的数据传输格式如图 3.17 所示。

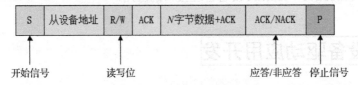

图 3.17 I2C 总线的数据传输格式

1）I2C 总线的通信过程

（1）主机首先发送开始信号，接着发送 1 B 的数据，该数据由高 7 位的地址码和最低 1 位的方向位组成（方向位表明主机与从机间数据的传输方向）。

（2）I2C 总线中所有的从机都对自己的地址与主机发送到总线上的地址进行比较，如果从机的地址与总线上的地址相同，则该从机就是要与主机进行数据传输的设备。

（3）传输数据，根据方向位，主机接收从机发送的数据或向从机发送数据。

（4）在数据传输完成后，主机发送一个停止信号，释放 I2C 总线。

（5）所有的从机等待下一个开始信号的到来。

2）I2C 总线的读写过程

主机向从机写数据的过程如图 3.18 所示。

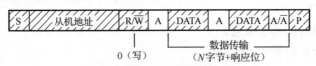

图 3.18 主机向从机写数据的过程

主机从从机读数据的过程如图 3.19 所示。

图 3.19 主机从从机读数据的过程

其中 S 表示由主机发送的开始信号，这时连接在 I2C 总线上的所有从机都会接收到开始信号。主机发送开始信号后，所有从机就开始等待主机接下来广播的从机地址信号（地址码）。在 I2C 总线上，每个设备的地址都是唯一的，当主机广播的地址与从机的地址相同时，这个从机就被选中了，没被选中的从机将会忽略之后传输的数据。根据 I2C 总线原理，这个从机地址可以是 7 位或 10 位的。在地址码之后，传输的是方向位，当该位为 0 时，表示后面的数据传输方向是由主机传输到从机，即主机向从机写数据；当该位为 1 时，传输方向则相反，即主机从从机读数据。从机接收到匹配的地址码后，主机或从机会返回一个应答（ACK）或非应答（NACK）信号，只有接收到 ACK 信号后，主机才能继续发送或接收数据。

写数据过程：主机在广播完地址码并接收到 ACK 信号后，开始向从机发送数据，数据包的大小为 8 位，主机每发送完 1 B 的数据后，都要等待从机的 ACK 信号。重复这个过程，主机可以向从机发送 N 个数据（N 没有大小限制）。当数据传输结束时，主机向从机发送一个停止信号（P），表示不再发送数据。

读数据过程：主机在广播完地址并接收到 ACK 信号后，从机开始向主机发送数据，数据包大小也是 8 位，从机每发送完 1 B 的数据，都会等待主机的 ACK 信号。重复这个过程，从机可以向主机发送 N 个数据，（N 没有大小限制）。当主机希望停止接收数据时，就向从机返回一个 ACK 信号，从机自动停止发送数据。

3.6.1.2　I2C 设备的管理方式

1）查找 I2C 设备

应用程序需要先根据 I2C 设备名称获取 I2C 设备句柄后，才能操作 I2C 设备。通过下面的函数可查找 I2C 设备：

```
/*******************************************************************
 * 名称：rt_device_find()
 * 功能：查找 I2C 设备
 * 参数：name 表示 I2C 设备名称
 * 返回：设备句柄表示查找到对应的 I2C 设备；RT_NULL 表示没有找到对应的 I2C 设备
 *******************************************************************/
rt_device_t rt_device_find(const char* name);
```

2）数据传输

获取到 I2C 设备句柄后就可以使用 rt_i2c_transfer()传输数据。代码如下：

```
/*******************************************************************
 * 名称：rt_i2c_transfer()
 * 功能：数据传输
 * 参数：bus 表示 I2C 设备句柄；msgs[]表示保存待传输消息的数组；num 表示消息数组的元素个数
 * 返回：消息数组的元素个数表示数据传输成功；错误码表示数据传输失败
 *******************************************************************/
rt_size_t rt_i2c_transfer(struct rt_i2c_bus_device   *bus, struct rt_i2c_msg msgs[], rt_uint32_t num);
```

3.6.2　开发设计与实践

3.6.2.1　硬件设计

本节通过 MobaXterm 串口终端 FinSH 控制台来获取传感器采集的数据，采用的是 I2C 总线。FinSH 控制台、ZI-ARMEmbed 和传感器的连接如图 3.20 所示。

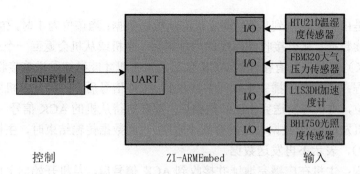

图 3.20   FinSH 控制台、ZI-ARMEmbed 和传感器

3.6.2.2   软件设计

软件设计流程如图 3.21 所示。

（1）在主函数中完成 I2C 设备的初始化。

（2）当在 MobaXterm 串口终端 FinSH 控制台中输入 "i2cDrv read" 命令时，可调用 read_temp()和 read_humi()函数可读取传感器采集的温度值和湿度值，并将其显示在 FinSH 控制台中。

（3）当在 MobaXterm 串口终端 FinSH 控制台中输入 "i2cDrv press" 命令时，可调用 fbm320_read_data()函数读取传感器采集的大气压强值，并将其显示在 FinSH 控制台中。

（4）当在 MobaXterm 串口终端 FinSH 控制台中输入 "i2cDrv acc" 命令时，可调用 lis3dh_read_data()函数读取传感器采集的加速度值，并将其显示在 FinSH 控制台中。

（5）当在 MobaXterm 串口终端 FinSH 控制台中输入 "i2cDrv light" 命令时，可调用 bh1750_get_data()函数读取传感器采集的光照度值，并将其显示在 FinSH 控制台中。

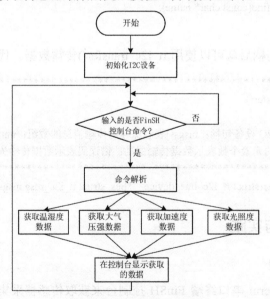

图 3.21   软件设计流程

3.6.2.3   功能设计与核心代码设计

要控制 I2C 设备，需要使用 RT-Thread 提供的 I2C 设备管理接口来初始化 I2C 设备。本节使用 FinSH 控制台对 I2C 设备进行控制，需要自定义 FinSH 控制台命令，并将已实现命

令添加到 FinSH 控制台命令列表中，这样在启动 MobaXterm 串口终端 FinSH 控制台后就可以使用自定义 FinSH 控制台命令了。

1）主函数（zonesion/app/main.c）

主函数的主要工作是完成 I2C 设备的初始化，在完成初始化后退出主函数。代码如下：

```
/**********************************************************************
 * 名称：main()
 * 功能：初始化 I2C 设备
 **********************************************************************/
#include "drv_hut21d/htu21d_drv.h"
#include "drv_pressure/fbm320_drv.h"
#include "drv_acc/lis3dh_drv.h"
#include "drv_bh1750/bh1750_drv.h"

int main(void)
{
    htu21d_init();                                  //初始化温湿度传感器
    fbm320_init();                                  //初始化大气压强传感器
    lis3dh_init();                                  //初始化加速度计
    bh1750_init();                                  //初始化光照度传感器

    return 0;
}
```

2）HTU21D 温湿度传感器初始化函数（zonesion/common/DRV/drv_hut21d/htu21d_drv.c）

HTU21D 温湿度传感器初始化函数的主要工作是对 HTU21D 温湿度传感器进行初始化。代码如下：

```
/**********************************************************************
 * 名称：htu21d_init()
 * 功能：初始化 HTU21D 温湿度传感器
 * 返回：0 表示初始化成功；-1 表示初始化失败
 **********************************************************************/
int htu21d_init(void)
{
    i2c_bus = (struct rt_i2c_bus_device *)rt_device_find(HTU21D_I2C_BUS_NAME);
    if(i2c_bus == RT_NULL)
    {
        rt_kprintf("failed to find %s device!\n", HTU21D_I2C_BUS_NAME);
        return -1;
    }
    write_reg(i2c_bus, HTU21D_RESET, 0);
    rt_thread_mdelay(50);
    return 0;
}
```

3）FBM320 大气压强传感器初始化函数（zonesion/common/DRV/drv_pressure/fbm320_drv.c）

FBM320大气压强传感器初始化函数的主要工作是对FBM320大气压强传感器进行初始化。代码如下：

```
/**********************************************************************************
 * 名称：fbm320_init (void)
 * 功能：初始化 FBM320 大气压强传感器
 * 返回：0 表示初始化成功；-1 表示初始化失败
 **********************************************************************************/
int fbm320_init(void)
{
    fbm320_bus = (struct rt_i2c_bus_device *)rt_device_find(FBM320_I2C_BUS_NAME);
    if(fbm320_bus == RT_NULL)
    {
        rt_kprintf("failed to find %s device!\n", FBM320_I2C_BUS_NAME);
        return -1;
    }
    unsigned char id = 0;
    fbm320_read_reg(fbm320_bus, FBM320_ID_ADDR, 1, &id);              //读取 FBM320 的 ID
    if(FBM320_ID != id)                                               //对比 ID
        return 1;
    Coefficient();
    return 0;
}
```

4）LIS3DH 加速度计初始化函数（zonesion/common/DRV/drv_acc/lis3dh_drv.c）

LIS3DH 加速度计初始化函数的主要工作是对 LIS3DH 加速度计进行初始化。代码如下：

```
/**********************************************************************************
 * 名称：lis3dh_init (void)
 * 功能：初始化 LIS3DH 加速度计
 **********************************************************************************/
int lis3dh_init(void)
{
    lis3dh_bus = (struct rt_i2c_bus_device *)rt_device_find(LIS3DH_I2C_BUS_NAME);
    if(lis3dh_bus == RT_NULL)
    {
        rt_kprintf("failed to find %s device!\n", LIS3DH_I2C_BUS_NAME);
        return -1;
    }
    unsigned char id = 0;
    lis3dh_read_reg(lis3dh_bus, LIS3DH_IDADDR, 1, &id);
    if(LIS3DH_ID != id)                                              //读取设备 ID
        return 1;
    rt_thread_mdelay(50);                                           //短延时
    unsigned char cmd = 0x97;
    if(lis3dh_write_reg(lis3dh_bus, LIS3DH_CTRL_REG1, &cmd, 2))      //1.25 kHz，使能三轴的输出
```

```
            return 1;
        rt_thread_mdelay(50);                                    //短延时
        cmd = 0x10;
        if(lis3dh_write_reg(lis3dh_bus, LIS3DH_CTRL_REG4, &cmd, 2))    //最大值为 4 倍的重力加速度
            return 1;
        return 0;
    }
```

5）BH1750 光照度传感器初始化函数（zonesion/common/DRV/drv_bh1750/bh1750_drv.c）

BH1750 光照度传感器初始化函数的主要工作是对 BH1750 光照度传感器进行初始化。
代码如下：

```
    /********************************************************************************
    * 名称：bh1750_init()
    * 功能：初始化 BH1750 光照度传感器
    ********************************************************************************/
    int bh1750_init(void)
    {
        i2c_bus = (struct rt_i2c_bus_device *)rt_device_find(BH1750_I2C_BUS_NAME);
        if(i2c_bus == RT_NULL)
        {
            rt_kprintf("failed to find %s device!\n", BH1750_I2C_BUS_NAME);
            return -1;
        }
        write_reg(i2c_bus, BH1750_POWER, 0);
        return 0;
    }
```

6）自定义 FinSH 控制台命令函数模块（zonesion/app/finsh_cmd/finsh_ex.c）

自定义 FinSH 控制台命令函数的主要工作是实现 I2C 设备的管理命令，可在 FinSH 控
制台中使用这些命令来读取 I2C 设备采集的数据。代码如下：

```
    /********************************************************************************
    * 名称：i2cDrv (int argc, char **argv)
    * 功能：自定义 FinSH 控制台命令
    * 参数：argc 表示参数个数；argv 表示指向参数字符串指针的数组指针
    ********************************************************************************/
    #include <rtthread.h>
    #include <stdio.h>
    #include "drv_hut21d/htu21d_drv.h"
    #include "drv_pressure/fbm320_drv.h"
    #include "drv_acc/lis3dh_drv.h"
    #include "drv_bh1750/bh1750_drv.h"
    int i2cDrv(int argc, char **argv)
    {
        float readdata[3];
        char tempBuf[64] = {0};
        if(!rt_strcmp(argv[1], "temp")) {
            //获取当前温湿度值并打包
            sprintf(tempBuf, "T = %.1f ℃, H = %.1f%%\n", read_temp(), read_humi());
```

```
            rt_kprintf(tempBuf);
        } else if(!rt_strcmp(argv[1], "press")) {
            //获取当前大气压强值并打包
            fbm320_read_data(readdata, readdata+1);
            sprintf(tempBuf, "Pressure = %.1fhPa\n", readdata[1]/100.f);
            rt_kprintf(tempBuf);
        } else if(!rt_strcmp(argv[1], "acc")) {
            //获取当前加速度值并打包
            lis3dh_read_data(readdata, readdata+1, readdata+2);
            sprintf(tempBuf, "acc[x] = %.1f, acc[y] = %.1f, acc[z] = %.1f\n", readdata[0], readdata[1],
readdata[2]);
            rt_kprintf(tempBuf);
        } else if(!rt_strcmp(argv[1], "light")) {
            //获取当前光照度值并打包
            readdata[0] = bh1750_get_data();
            sprintf(tempBuf, "Light = %.1flux\n", readdata[0]);
            rt_kprintf(tempBuf);
        }
        return 0;
    }
MSH_CMD_EXPORT(i2cDrv, i2c driver sample);
```

3.6.3　开发步骤与验证

3.6.3.1　硬件部署

同 1.2.3.1 节。

3.6.3.2　工程调试

（1）将本项目工程文件夹复制到 RT-ThreadStudio\workspace 工作区目录下。

（2）其余同 2.1.3.2 节。

3.6.3.3　验证效果

（1）关闭 RT-Thread Studio，拔掉仿真器，按下 ZI-ARMEmbed 上的电源按键重新上电。

（2）在 MobaXterm 串口终端 FinSH 控制台中输入"i2cDrv temp"命令，解析自定义 FinSH 控制台命令后可采集温湿度值，并在 FinSH 控制台输出 "T = 27.8 ℃, H = 32.0%"。

```
[I/I2C] I2C bus [i2c1] registered

 \ | /
- RT -      Thread Operating System
 / | \      4.1.0 build Oct 24 2022 11:48:03
 2006 - 2022 Copyright by RT-Thread team
msh >i2cDrv temp
T = 27.8 ℃, H = 32.0%
msh >
```

（3）在 MobaXterm 串口终端 FinSH 控制台中输入"i2cDrv press"命令，解析自定义 FinSH 控制台命令后可采集大气压强值，并在 FinSH 控制台输出 "Pressure = 1012.8hPa"。

```
[I/I2C] I2C bus [i2c1] registered

 \ | /
- RT -     Thread Operating System
 / | \     4.1.0 build Oct 24 2022 11:48:03
 2006 - 2022 Copyright by RT-Thread team
msh >i2cDrv temp
T = 27.8 ℃, H = 32.0%
msh >i2cDrv press
Pressure = 1012.8hPa
msh >
```

（4）在 MobaXterm 串口终端 FinSH 控制台中输入"i2cDrv acc"命令，解析自定义 FinSH 控制台命令后可采集加速度值，并在 FinSH 控制台输出"acc[x] = 0.2, acc[y] = -0.2, acc[z] = 9.8"。调整 ZI-ARMEmbed 的角度，重新输入"i2cDrv acc"命令，加速度值会发生变化。

```
[I/I2C] I2C bus [i2c1] registered

 \ | /
- RT -     Thread Operating System
 / | \     4.1.0 build Oct 24 2022 11:48:03
 2006 - 2022 Copyright by RT-Thread team
msh >i2cDrv temp
T = 27.8 ℃, H = 32.0%
msh >i2cDrv press
Pressure = -260.8hPa
msh >i2cDrv acc
acc[x] = 0.2, acc[y] = -0.2, acc[z] = 9.8
msh >
```

（5）在 MobaXterm 串口终端 FinSH 控制台中输入"i2cDrv light"命令，解析自定义 FinSH 控制台命令后可采集加速度值，并在 FinSH 控制台输出"Light = 67.5lux"。打开手机的手电筒，在照射光照传感器时，重新输入"i2cDrv light"命令，光照度值会发生变化。

```
[I/I2C] I2C bus [i2c1] registered

 \ | /
- RT -     Thread Operating System
 / | \     4.1.0 build Oct 24 2022 11:48:03
 2006 - 2022 Copyright by RT-Thread team
msh >i2cDrv temp
T = 27.8 ℃, H = 32.0%
msh >i2cDrv press
Pressure = -260.8hPa
msh >i2cDrv acc
acc[x] = 0.2, acc[y] = -0.2, acc[z] = 9.8
msh >i2cDrv light
Light = 67.5lux
msh >
```

3.6.4　小结

本节主要介绍 I2C 设备的基本概念、通信过程和管理方式。通过本节的学习，读者可掌握 RT-Thread I2C 设备管理接口的应用。

3.7 PWM 设备驱动应用开发

脉冲宽度调制（Pulse Width Modulation，PWM）是一种通过调整信号的脉冲宽度来控制平均功率的技术，广泛应用于电子和通信领域，常见的应用包括电机控制、LED 亮度调节、音频调制等。PWM 是通过快速、周期性地切换信号的高和低电平来实现的。PWM 的应用领域有：

- 电机控制：PWM 广泛应用于电机转动速度和方向的控制，通过调整信号的占空比可调整电机的平均功率。
- LED 亮度调节：通过调整 PWM 信号的占空比可调整 LED 亮度。
- 电源调节：在开关电源中，通过调整 PWM 信号的占空比可调整输出电压。
- 音频调制：在数字音频处理中，通过控制 PWM 信号的脉冲宽度可调整音量。

本节的要求如下：

- 了解 PWM 设备的基本概念。
- 掌握 PWM 设备的管理方式。
- 掌握 RT-Thread PWM 设备管理接口的应用。

3.7.1　原理分析

3.7.1.1　PWM 简介

PWM 原理图如图 3.22 所示，假设定时器的工作模式为向上计数，当计数值小于阈值时，则输出一种电平，如高电平；当计数值大于阈值时则输出相反的电平，如低电平。当计数值达到最大值时，定时器从 0 开始重新计数，又回到最初的电平。高电平的持续时间（脉冲宽度）和周期时间的比值就是占空比，其范围为 0%～100%。

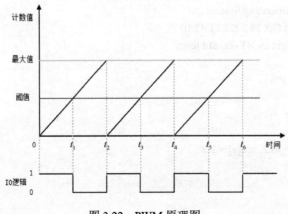

图 3.22　PWM 原理图

3.7.1.2 访问 PWM 设备

1）查找 PWM 设备

应用程序可根据 PWM 设备名称获取设备句柄，进而操作 PWM 设备。通过下面的函数可查找 PWM 设备：

```
/**************************************************************************
 * 名称：rt_device_find()
 * 功能：查找 PWM 设备
 * 参数：name 表示 PWM 设备名称
 * 返回：PWM 设备句柄表示查找到对应 PWM 设备；RT_NULL 表示没有找到 PWM 设备
 **************************************************************************/
rt_device_t rt_device_find(const char* name);
```

2）设置 PWM 设备的周期和脉冲宽度

通过下面的函数可设置 PWM 设备的周期和脉冲宽度：

```
/**************************************************************************
 * 名称：rt_pwm_set()
 * 功能：设置 PWM 设备的周期和脉冲宽度
 * 参数：device 表示 PWM 设备句柄；channel 表示 PWM 设备通道；period 表示 PWM 设备的周期（单
位为 ns，当 period 为 500000 时，表示周期为 0.5 ms）；pulse 表示 PWM 设备的脉冲宽度（单位为 ns，pulse
不能超过 period）
 * 返回：RT_EOK 表示成功；-RT_EIO 表示 PWM 设备为空；-RT_ENOSYS 表示 PWM 设备操作方
法为空；其他错误码表示执行失败
 **************************************************************************/
rt_err_t rt_pwm_set(struct rt_device_pwm *device, int channel, rt_uint32_t period, rt_uint32_t pulse);
```

3）使能 PWM 设备

设置 PWM 设备的周期和脉冲宽度后，通过下面的函数可使能 PWM 设备：

```
/**************************************************************************
 * 名称：rt_pwm_enable()
 * 功能：使能 PWM 设备
 * 参数：device 表示 PWM 设备句柄；channel 表示 PWM 设备通道；
 * 返回：RT_EOK 表示 PWM 设备使能成功；-RT_ENOSYS 表示 PWM 设备操作方法为空；其他错
误码表示 PWM 设备使能失败
 **************************************************************************/
rt_err_t rt_pwm_enable(struct rt_device_pwm *device, int channel);
```

4）关闭 PWM 设备通道

通过下面的函数可关闭 PWM 设备通道：

```
/**************************************************************************
 * 名称：rt_pwm_disable()
 * 功能：关闭 PWM 设备对应通道
 * 参数：device 表示 PWM 设备句柄；channel 表示 PWM 设备通道
 * 返回：RT_EOK 表示 PWM 设备关闭成功；-RT_EIO 表示 PWM 设备句柄为空；其他错误码表示
PWM 设备关闭失败
 **************************************************************************/
```

rt_err_t rt_rt_pwm_disable (struct rt_device_pwm *device, int channel);

3.7.2 开发设计与实践

3.7.2.1 硬件设计

本节通过 MobaXterm 串口终端 FinSH 控制台来控制 PWM 设备（如电机），目的是帮助读者掌握 PWM 设备管理接口的应用。FinSH、ZI-ARMEmbed 与电机的连接如图 3.23 所示。

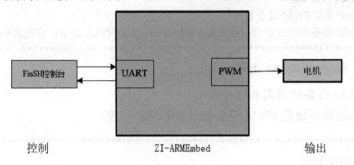

图 3.23　硬件设计

本节采用 A3967SLB 芯片（电机驱动器）来驱动电机，该芯片的硬件接口电路如图 3.24 所示。图中，A3967SLB 芯片的 nENABLE 引脚与 STM32F407VET6 的 PC9 引脚相连，A3967SLB 芯片的 DIR 引脚与 STM32F407VET6 的 PE4 相连。

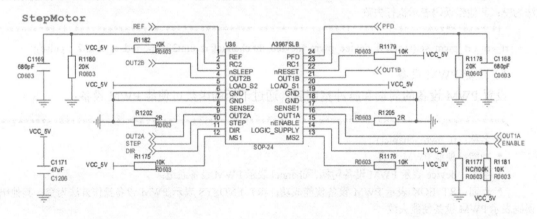

图 3.24　A3967SLB 芯片的硬件接口电路

3.7.2.2 软件设计

软件设计流程如图 3.25 所示。

（1）在主函数中对电机进行初始化。

（2）编写电机的控制函数，根据不同参数实现不同的动作。

（3）在 MobaXterm 串口终端 FinSH 控制台中输入相关控制命令，通过控制命令来调用对应的电机控制函数。

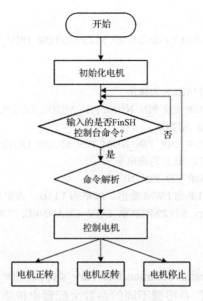

图 3.25　软件设计流程

3.7.2.3　功能设计与核心代码设计

要控制 PWM 设备，首先需要使用 RT-Thread 提供的 PWM 设备管理接口来完成对 PWM 设备的初始化，然后编写相应的控制函数。本节通过 FinSH 控制台命令来控制电机，需要自定义 FinSH 控制台命令，并将相应的命令添加到 FinSH 控制台命令列表中。

1) 主函数（zonesion/app/main.c）

主函数的主要工作是初始化电机，在初始化完成后退出主函数。代码如下：

```
/*********************************************************************************
* 名称：main()
* 功能：调用电机初始化函数
*********************************************************************************/
#include "drv_stepper/drv_stepMotor.h"
int main(void)
{
    drv_stepMotor_init();
    return 0;
}
```

2) 电机初始化函数（zonesion/common/DRV/drv_stepper/drv_stepMotor.c）

电机初始化函数的主要工作是查找 PWM 设备（如电机），配置相应的引脚，设置 PWM 设备通道、周期及脉冲宽度。代码如下：

```
/*********************************************************************************
* 名称：drv_stepMotor_init()
* 功能：电机初始化
*********************************************************************************/
int drv_stepMotor_init(void)
{
    stepMotorDrv = (struct rt_device_pwm *)rt_device_find(STEPMOTOR_DRV_NAME);//查找 PWM 设备
    if(stepMotorDrv == RT_NULL)
```

```
    {
        rt_kprintf("failed to find %s device!\n", STEPMOTOR_DRV_NAME);
        return -1;
    }
    //设置电机驱动器的 nENABLE 为输出
    rt_pin_mode(STEPMOTOR_EN_PIN_NUM, PIN_MODE_OUTPUT);
    //设置电机驱动器的 DIR 为输出
    rt_pin_mode(STEPMOTOR_DIR_PIN_NUM, PIN_MODE_OUTPUT);
    //设置电机驱动器的 nENABLE 为高电平
    rt_pin_write(STEPMOTOR_EN_PIN_NUM, 1);
    //设置电机驱动器的 STEP 为 PWM 输出、频率为 1 kHz、占空比为 50%
    rt_pwm_set(stepMotorDrv, STEPMOTOR_DRV_CHANNEL, 1000000, 500000);
    return 0;
}
```

3）电机控制函数（zonesion/common/DRV/drv_stepper/drv_stepMotor.c）

电机控制函数的主要工作是根据不同的参数来配置电机驱动器的 DIR 引脚和 nENABLE 引脚，然后使能电机，从而实现电机的正转、反转、停止等。代码如下：

```
/*****************************************************************************
* 名称：drv_stepMotor_ctrl()
* 功能：控制电机
* 参数：0 表示进电机停止；1 表示电机正转；其他数字表示电机反转
*****************************************************************************/
void drv_stepMotor_ctrl(unsigned char dir)
{
    if(dir == 0)
    {
        rt_pwm_disable(stepMotorDrv, STEPMOTOR_DRV_CHANNEL);          //失能电机
        rt_pin_write(STEPMOTOR_EN_PIN_NUM, 1);//设置电机驱动器的 nENABLE 为高电平，失能
    }
    else
    {
        if(dir == 1)
            rt_pin_write(STEPMOTOR_DIR_PIN_NUM, 0);        //设置电机驱动器的 DIR 为低电平
        else
            rt_pin_write(STEPMOTOR_DIR_PIN_NUM, 1);        //设置电机驱动器的 DIR 为高电平
        rt_pin_write(STEPMOTOR_EN_PIN_NUM, 0);//设置电机驱动器的 nENABLE 为低电平，使能
        rt_pwm_enable(stepMotorDrv, STEPMOTOR_DRV_CHANNEL);          //使能电机
    }
}
```

4）自定义 FinSH 控制台命令函数（zonesion/app/finsh_cmd/finsh_ex.c）

自定义 FinSH 控制台命令函数的主要工作是实现 PWM 设备的管理命令，可在 FinSH 控制台中使用这些命令来控制 PWM 设备。代码如下：

```
/*****************************************************************************
* 名称：stepmotor()
* 功能：自定义电机的 FinSH 控制台控制命令
```

```
* 参数：argc 表示参数个数；argv 表示指向参数字符串指针的数组指针；argv[0]为 stepmotor 命令；argv[1]
为命令后的参数
*************************************************************************/
int stepmotor(int argc, char **argv)
{
    if(argc == 2)
    {
        drv_stepMotor_ctrl(atoi(argv[1]));              //调用电机控制函数 drv_stepMotor_ctrl()
    }
    else
        rt_kprintf("Please input 'stepmotor <dir(0-stop 1-> 2-<)>'\n");
    return 0;
}
MSH_CMD_EXPORT(stepmotor, stepmotor driver sample);   //将 stepmotor 命令添加到 FinSH 控制台命
令列表中
```

3.7.3　开发步骤与验证

3.7.3.1　硬件部署

同 1.2.3.1 节。

3.7.3.2　工程调试

（1）将本项目工程文件夹复制到 RT-ThreadStudio\workspace 工作区目录下。

（2）其余同 2.1.3.2 节。

3.7.3.3　验证效果

（1）关闭 RT-Thread Studio，拔掉仿真器，按下 ZI-ARMEmbed 上的电源按键重新上电。

（2）在 MobaXterm 串口终端 FinSH 控制台中输入"stepmotor 1"命令，解析自定义 FinSH 控制台命令后可实现电机的正转。ZI-ARMEmbed 中的电机如图 3.26 所示。

```
 \ | /
- RT -     Thread Operating System
 / | \     4.1.0 build Oct 24 2022 14:49:15
 2006 - 2022 Copyright by RT-Thread team
msh >stepmotor 1
msh >
```

图 3.26　ZI-ARMEmbed 中的电机

（3）在 MobaXterm 串口终端 FinSH 控制台中输入"stepmotor 2"命令，解析自定义 FinSH 控制台命令后可实现电机的反转。

```
  \ | /
- RT -     Thread Operating System
 / | \     4.1.0 build Oct 24 2022 14:49:15
 2006 - 2022 Copyright by RT-Thread team
msh >stepmotor 1
msh >stepmotor 2
msh >
```

（4）在 MobaXterm 串口终端 FinSH 控制台中输入"stepmotor 0"命令，解析自定义 FinSH 控制台命令后可停止电机。

```
  \ | /
- RT -     Thread Operating System
 / | \     4.1.0 build Oct 24 2022 14:49:15
 2006 - 2022 Copyright by RT-Thread team
msh >stepmotor 1
msh >stepmotor 2
msh >stepmotor 0
msh >
```

3.7.4　小结

本节主要介绍 PWM 设备的基本概念和管理方式。通过本节的学习，读者可掌握 RT-Thread PWM 设备管理接口的应用。

3.8 RTC 设备驱动应用开发

实时时钟（Real-Time Clock，RTC）是一种专门用于在计算机系统和嵌入式系统中提供实时时间的设备或模块。RTC 通常包含一个稳定的时钟源。即使计算机系统和嵌入式系统处于关闭状态或低功耗模式，RTC 也可以持续追踪当前时间。RTC 的应用领域有：

- ⊃ 计算机系统：在计算机系统中，RTC 用于维护系统时钟，并在需要时唤醒系统。
- ⊃ 嵌入式系统：在嵌入式系统中，RTC 可以用于记录时间戳、定时操作等。
- ⊃ 实时数据记录：RTC 可用于实时数据的记录，确保数据被准确的时间戳记录。

本节的要求如下：

- ⊃ 了解 RTC 的基本概念。
- ⊃ 掌握 RTC 设备的管理方式。
- ⊃ 掌握 RT-Thread RTC 设备管理接口的应用。

3.8.1　原理分析

3.8.1.1　RTC 简介

RTC 可提供精确的实时时间，用于产生年、月、日、时、分、秒等信息。目前 RTC 设备大多采用精度较高的晶体振荡器作为时钟源。为了使 RTC 设备在主电源掉电时还能工作，会外加电池供电，从而使时间信息保持有效。

RT-Thread 的 RTC 设备提供了基础的时间服务。面对越来越多的 IoT 场景，RTC 设备已经成为产品的标配，甚至在诸如 SSL 等的安全传输过程中，RTC 设备也成了不可或缺的组成部分。RTC 设备的特点如下：

（1）独立时钟源。RTC 设备通常有独立的时钟源，如晶体振荡器或其他稳定的时钟源。这使得 RTC 设备能够在计算机系统和嵌入式系统处于关闭状态或低功耗模式的情况下持续运行。

（2）时间保持。RTC 设备的主要任务是追踪当前时间，即使在计算机系统或嵌入式系统断电或重启后，RTC 设备也能够保持精确的时间。

（3）备用电池。RTC 设备通常使用备用电池，即使在主电源关闭时，RTC 设备仍能维持时间记录。

（4）系统唤醒。RTC 设备通常具有闹钟或定时器功能，可以用来唤醒计算机系统或嵌入式系统。这对于需要在特定时间执行任务且需要保持低功耗状态的应用来说，是非常重要的。

3.8.1.2　访问 RTC 设备

1）设置 RTC 设备的当前日期

通过下面的函数可设置 RTC 设备的当前日期：

```
/*****************************************************************************
 * 名称：set_date()
 * 功能：设置 RTC 设备当前日期
 * 参数：year 表示待设置生效的年份；month 表示待设置生效的月份；day 表示待设置生效的日
 * 返回：RT_EOK 表示设置成功；-RT_ERROR 表示失败（没有找到 RTC 设备）；其他错误码表示
失败
 *****************************************************************************/
rt_err_t set_date(rt_uint32_t year, rt_uint32_t month, rt_uint32_t day)
```

2）设置 RTC 设备的当前时间

通过下面的函数可设置 RTC 设备的当前时间：

```
/*****************************************************************************
 * 名称：set_time()
 * 功能：设置 RTC 设备当前时间
 * 参数：hour 表示待设置生效的小时；minute 表示待设置生效的分钟；second 表示待设置生效的秒
 * 返回：RT_EOK 表示设置成功；-RT_ERROR 表示失败（没有找到 RTC 设备）；其他错误码表示失败
 *****************************************************************************/
rt_err_t set_time(rt_uint32_t hour, rt_uint32_t minute, rt_uint32_t second)
```

3）获取 RTC 设备的当前时间

通过下面的函数可获取 RTC 设备的当前时间：

```
/*****************************************************************************
 * 名称：time()
 * 功能：获取 RTC 设备当前时间
 * 参数：t 表示指向 RTC 设备当前时间的指针
 * 返回：当前时间值
 *****************************************************************************/
time_t time(time_t *t)
```

4）配置 RTC 设备的功能

（1）启用 Soft RTC 功能。RT-Thread 可通过软件来模拟 RTC 设备。在 menuconfig 中可以配置 Soft RTC 功能，该功能非常适用于对时间精度要求不高、没有 RTC 设备的产品。配置代码如下：

```
RT-Thread Components
    Device Drivers:
        -*- Using RTC device drivers                              //使用 RTC 设备驱动
        [ ]    Using software simulation RTC device               //启用 Soft RTC 功能
```

（2）使用网络时间协议（Network Time Protocol，NTP）来实现时间自动同步。如果 RT-Thread 已接入互联网，可使用 NTP 来实现时间自动同步，从而定期同步本地时间。该功能同样也可以在 menuconfig 中配置。配置代码如下：

```
RT-Thread online packages
    IoT - internet of things
        netutils: Networking utilities for RT-Thread:
            [*]    Enable NTP(Network Time Protocol) client
```

开启 RTC 设备的时间自动同步功能后，可以设置同步周期和首次同步的延时时间：

```
RT-Thread Components →
    Device Drivers:
        -*- Using RTC device drivers                              //使用 RTC 设备驱动
        [ ]    Using software simulation RTC device               //启用 Soft RTC 功能
        [*]    Using NTP auto sync RTC time                       //使用 NTP 实现时间自动同步
        (30)    NTP first sync delay time(second) for network connect    /*首次使用 NTP 实现时间
自动同步的延时。延时的目的是给网络连接预留一定的时间，提高第一次使用 NTP 实现时间自动同步时的
成功率。默认的延时为 30 s*/
        (3600)  NTP auto sync period(second)                      /*使用 NTP 实现时间自动同步的
周期，单位为 s，默认的周期为每小时同步一次。*/
```

3.8.2 开发设计与实践

3.8.2.1 硬件设计

本节通过 MobaXterm 串口终端 FinSH 控制台来控制 RTC 设备，帮助读者掌握 RTC 设备管理接口的应用。FinSH 控制台和 ZI-ARMEmbed 的连接如图 3.27 所示。

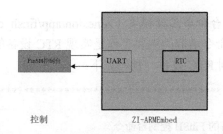

图 3.27　FinSH 控制台和 ZI-ARMEmbed 的连接

3.8.2.2　软件设计

软件设计流程如图 3.28 所示。

（1）启动系统后显示 RTC 设备的当前日期和时间。

（2）在 MobaXterm 串口终端 FinSH 控制台中输入控制命令来获取 RTC 设备的当前日期和时间，并设置 RTC 设备的当前日期和时间。

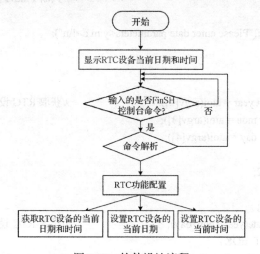

图 3.28　软件设计流程

3.8.2.3　功能设计与核心代码设计

1）主函数（zonesion/app/main.c）

主函数的主要工作是调用 RTC 设备管理的接口，获取并显示 RTC 设备的当前日期和时间。代码如下：

```
/*******************************************************************************
* 名称：main()
* 功能：获取并显示 RTC 设备的当前日期和时间
*******************************************************************************/
#include <rtdevice.h>
int main(void)
{
    time_t now;
    now = time(RT_NULL);                          //获取并显示 RTC 设备的当前日期和时间
    rt_kprintf("%s\n", ctime(&now));
    return 0;
}
```

2）自定义 FinSH 控制台命令函数模块（zonesion/app/finsh_cmd/finsh_ex.c）

自定义 FinSH 控制台命令函数的主要工作是实现 RTC 设备的管理命令，可在 FinSH 控制台中使用这些命令来控制 RTC 设备。代码如下：

```
/*******************************************************************************
* 名称：rtcDrv()
* 功能：自定义 RTC 设备的 FinSH 控制台命令
* 参数：argc 表示参数个数；argv 表示指向参数字符串指针的数组指针
*******************************************************************************/
int rtcDrv(int argc, char **argv)
{
    rt_err_t result = RT_EOK;
    if(!rt_strcmp(argv[1], "setDate"))
    {
        if(argc < 5)                                   //判断 FinSH 控制台命令是否带有参数
        {
            rt_kprintf("Please enter date parameters<y m d>!\n");
            return 0;
        }

        unsigned short year = atoi(argv[2]);           //获取 RTC 设备的当前日期
        unsigned char mon = atoi(argv[3]);
        unsigned char day = atoi(argv[4]);
        if(mon > 12)
            mon = 12;
        if(day > 31)
            day = 31;
        result = set_date(year, mon, day);             //设置 RTC 设备的当前日期
        if(result == RT_EOK)
            rt_kprintf("set date: %d.%d.%d\n", year, mon, day);
        else
            rt_kprintf("setting date failed!\n");
    }
    else if(!rt_strcmp(argv[1], "setTime"))
    {
        if(argc < 5)                                   //判断 FinSH 控制台命令是否带有参数
        {
            rt_kprintf("Please enter time parameters<h m s>!\n");
            return 0;
        }
        unsigned char hour = atoi(argv[2]);            //获取 RTC 设备的当前时间
        unsigned char min = atoi(argv[3]);
        unsigned char sec = atoi(argv[4]);
        if(hour >= 24)
            hour = 0;
        if(min >= 60)
            min = 0;
```

```
        if(sec >= 60)
            sec = 0;
        result = set_time(hour, min, sec);                    //设置 RTC 设备的当前时间
        if(result == RT_EOK)
            rt_kprintf("set time: %d:%d:%d\n", hour, min, sec);
        else
            rt_kprintf("setting time failed!\n");
    }
    else if(!rt_strcmp(argv[1], "get"))
    {
        time_t now;
        now = time(RT_NULL);                                  //获取 RTC 设备的当前日期和时间
        rt_kprintf("date: %s\n", ctime(&now));
    }
    else
        rt_kprintf("Please input 'rtcDrv <get|setDate|setTime>'\n");
    return 0;
}
MSH_CMD_EXPORT(rtcDrv, rtc driver sample);                    //将 rtcDrv 命令添加到 FinSH 控制台命令列表中
```

3.8.3　开发步骤与验证

3.8.3.1　硬件部署

同 1.2.3.1 节。

3.8.3.2　项目部署

（1）将本项目工程文件夹复制到 RT-ThreadStudio\workspace 工作区目录下。

（2）部署公共文件（公共文件位于本书配套资源的"02-软件资料\01-操作系统\rtt-common.zip"中）：将 rt-thread 文件夹复制到工程的根目录，将 zonesion\common 文件夹复制到工程的 zonesion 目录下。注意：本节需要注释掉"rt-thread\components\drivers\include\drivers\rtc.h"文件中的 rt_hw_rtc_register 声明函数，否则会出现函数定义重复错误）。

（3）使用 RT-Thread Studio 导入本项目软件包。

3.8.3.3　验证效果

（1）关闭 RT-Thread Studio，拔掉仿真器，按下 ZI-ARMEmbed 上的电源按键重新上电。

（2）在 MobaXterm 串口终端 FinSH 控制台中输入命令"rtcDrv get"，解析自定义 FinSH 控制台命令后可获取 RTC 设备的当前日期和时间，并在 FinSH 控制台输出 RTC 设备的当前日期和时间。

```
 \ | /
- RT -     Thread Operating System
 / | \     4.1.0 build Oct 24 2022 17:31:01
 2006 - 2022 Copyright by RT-Thread team
Mon Oct 24 17:40:22 2022


msh >rtcDrv get
```

```
date: Mon Oct 24 17:40:51 2022

msh >
```

（3）在 MobaXterm 串口终端 FinSH 控制台中输入命令"rtcDrv setDate 2022 10 24"，解析自定义 FinSH 控制台命令后可设置 RTC 设备的当前日期，这里将 RTC 设备的当前日期设置为 2022 年 10 月 24 日。

```
 \ | /
- RT -      Thread Operating System
 / | \      4.1.0 build Oct 24 2022 17:31:01
 2006 - 2022 Copyright by RT-Thread team
Mon Oct 24 17:40:22 2022

msh >rtcDrv get
date: Mon Oct 24 17:40:51 2022

msh >rtcDrv setDate 2022 10 24
set date: 2022.10.24
msh >
```

（4）在 MobaXterm 串口终端 FinSH 控制台中输入命令"rtcDrv setTime 17 42 45"，解析自定义 FinSH 控制台命令后可设置 RTC 设备的当前时间，这里将 RTC 设备的当前时间设置为 17 时 42 分 45 秒。

```
 \ | /
- RT -      Thread Operating System
 / | \      4.1.0 build Oct 24 2022 17:31:01
 2006 - 2022 Copyright by RT-Thread team
Mon Oct 24 17:40:22 2022

msh >rtcDrv get
date: Mon Oct 24 17:40:51 2022

msh >rtcDrv setDate 2022 10 24
set date: 2022.10.24
msh >rtcDrv setTime 17 42 45
set time: 17:42:45
msh >
```

（5）在 MobaXterm 串口终端 FinSH 控制台中输入命令"rtcDrv get"，解析自定义 FinSH 控制台命令后，FinSH 控制台将输出 RTC 设备的当前日期和时间。

```
 \ | /
- RT -      Thread Operating System
 / | \      4.1.0 build Oct 24 2022 17:31:01
 2006 - 2022 Copyright by RT-Thread team
Mon Oct 24 17:40:22 2022
```

```
msh >rtcDrv get
date: Mon Oct 24 17:40:51 2022

msh >rtcDrv setDate 2022 10 24
set date: 2022.10.24
msh >rtcDrv setTime 17 42 45
set time: 17:42:45
msh >rtcDrv get
date: Mon Oct 24 17:42:51 2022

msh >
```

3.8.4　小结

本节主要介绍 RTC 设备的基本概念和管理方式。通过本节的学习，读者可掌握 RT-Thread RTC 设备管理接口的应用。

3.9 SPI 设备驱动应用开发

SPI（Serial Peripheral Interface）总线是一种用于串行同步通信的协议，通常用于在微控制器、传感器、存储器和其他外部设备之间传输数据。

本节的要求如下：

- 了解 SPI 设备的工作原理。
- 掌握 SPI 设备的管理方式。
- 掌握 RT-Thread SPI 设备管理接口的应用。

3.9.1　原理分析

3.9.1.1　SPI 简介

SPI 总线是一种高速全双工的通信总线，被广泛用于 ADC、LCD 等设备与 MCU 之间的、要求通信速率较高的场合。SPI 使用 4 根信号进行通信，分别是串行时钟线（SCK）、主出从入线（MOSI）、主入从出线（MISO）和片选线（CS）。SPI 主机和从机的连接如图 3.29 所示。

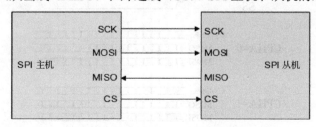

图 3.29　SPI 主机和从机的连接

SPI 总线的 4 根信号线的定义如下：

◯ MOSI：主出从入线（SPI Bus Master Output/Slave Input）。

◯ MISO：主入从出线（SPI Bus Master Input/Slave Output）。

◯ SCK：串行时钟线（Serial Clock），主机将时钟信号发送到从机。

◯ CS：片选线（Chip Select），主机将片选信号发送到从机。

SPI 总线采用主从工作方式，通常有一个主机和一个或多个从机。通信由主机发起，主机通过 CS 选择要通信的从机，然后通过 SCK 向从机发送时钟信号，数据通过 MOSI 发送到从机，同时通过 MISO 接收从机发送的数据。

SPI 总线的工作原理如图 3.30 所示，图中的 MCU 有 2 个 SPI 控制器，SPI 控制器相当于 SPI 主机，每个 SPI 主机可以连接多个 SPI 从机。挂载在同一个 SPI 主机上的从机可共享 SPI 控制器的 3 个引脚，即 SCK、MISO、MOSI 引脚，但每个 SPI 从机的 CS 引脚是独立的。SPI 主机通过 CS 引脚选择 SPI 从机，CS 引脚一般是低电平有效的。任何时刻，SPI 主机上只有一个 CS 引脚处于有效状态，与该 CS 引脚连接的 SPI 从机可与 SPI 主机通信。

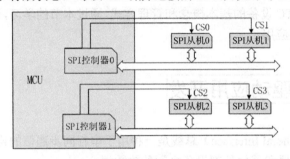

图 3.30   SPI 总线的工作原理

SPI 从机的时钟信号由 SPI 主机通过 SCK 提供，MOSI、MISO 基于时钟信号完成数据传输。SPI 总线的工作模式由 CPOL（Clock Polarity，时钟极性）和 CPHA（Clock Phase，时钟相位）共同决定，CPOL 是时钟信号的初始电平的状态，CPOL 为 0 表示时钟信号初始状态为低电平，为 1 表示时钟信号的初始电平是高电平。CPHA 表示在哪个时钟沿采样数据，CPHA 为 0 表示在首个时钟变化沿采样数据，而 CPHA 为 1 则表示在第二个时钟变化沿采样数据。根据 CPOL 和 CPHA 的不同组合，SPI 总线有 4 种工作模式：CPOL=0、CPHA=0，CPOL=0、CPHA=1，CPOL=1、CPHA=0，CPOL=1、CPHA=1。CPOL 和 CPHA 的工作时序如图 3.31 所示。

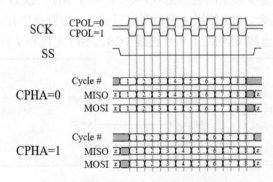

图 3.31   CPOL 和 CPHA 的工作时序

SPI 可分为标准 SPI、Dual SPI 和 Quad SPI。在相同时钟信号下，信号线越多传输速率越高。

（1）QSPI。QSPI 是 Queued SPI 的简写，是 SPI 的扩展，比 SPI 的应用更加广泛。QSPI 在 SPI 的基础上进行了功能增强，增加了队列传输机制。使用 QSPI，可以一次性传输包含 16 个 8 bit 或 16 bit 数据的传输队列。QSPI 允许使用 DMA 控制器来管理 SPI 的数据传输，无须 CPU 的干预，减轻了 CPU 的负担，提高了数据传输效率。与 SPI 相比，QSPI 的最大特点是用 80 B 的 RAM 代替了 SPI 的发送和接收数据寄存器。

（2）Dual SPI。通过发送一个命令字节，可以使 SPI 进入 Dual 模式，让它工作在半双工模式，从而使数据传输速率变为原来的 2 倍。在 Dual 模式下，MOSI 变成 SIO0（Serial IO 0），MISO 变成 SIO1（Serial IO 1），因此在一个时钟信号周期内就能传输 2 bit 的数据，加倍了数据传输速率。

（3）Quad SPI。Quad SPI 是一种改进型 SPI，可以实现更快的数据传输速率。Quad SPI 通常具有更大的存储容量，可用于更复杂的应用，如嵌入式操作系统和大型数据存储。

由 CPOL 和 CPHA 的不同状态，SPI 的工作模式可分为 4 种，如表 3.3 所示，API 主机与 API 从机的工作模式相同才能正常通信，多数的应用采用的是工作模式 0 与工作模式 3。

表 3.3　SPI 的工作模式

SPI 工作模式	CPOL 电平	CPHA 电平	SCK	采样时刻
工作模式 0	0	0	低电平	基数跳变沿
工作模式 1	0	1	低电平	偶数跳变沿
工作模式 2	1	0	高电平	基数跳变沿
工作模式 3	1	1	高电平	偶数跳变沿

3.9.1.2　SPI 相关配置

1）挂载 SPI 设备

通过下面的函数可将 SPI 设备挂载到 SPI 总线上：

```
/*******************************************************************************
 * 名称：rt_spi_bus_attach_device()
 * 功能：挂载 SPI 设备
 * 参数：device 表示 SPI 设备句柄；name 表示 SPI 设备名称；bus_name 表示 SPI 总线名称；user_data
表示指向 SPI 设备 CS 引脚的指针
 * 返回：RT_EOK 表示成功；其他错误码表示失败
 *******************************************************************************/
rt_err_t rt_spi_bus_attach_device(struct rt_spi_device *device, const char *name, const char *bus_name,
                                                          void *user_data)
```

函数 rt_spi_bus_attach_device() 用于将 SPI 设备挂载到指定的 SPI 总线上，向 RT-Thread 注册 SPI 设备，将 user_data 保存到 SPI 设备控制块中。SPI 总线命名原则为 spix，SPI 设备命名原则为 spixy，如 spi10 表示挂载在 spi1 总线上的 0 号设备。user_data 一般为指向 SPI 设备 CS 引脚的指针，在进行数据传输时 SPI 控制器会通过 CS 引脚进行片选。

2）配置 SPI 设备

通过下面的函数可配置 SPI 设备：

```
/**************************************************************************
 * 名称：rt_spi_configure()
 * 功能：配置 SPI 设备
 * 参数：device 表示 SPI 设备句柄；cfg 表示指向 SPI 设备配置参数的指针
 * 返回：RT_EOK 表示成功
 **************************************************************************/
rt_err_t rt_spi_configure(struct rt_spi_device *device, struct rt_spi_configuration *cfg)
```

3.9.1.3 访问 SPI 设备

一般情况，MCU 中的 SPI 设备通常是作为主机来和从机通信的，RT-Thread 将 SPI 主机虚拟为 SPI 设备，应用程序可使用 SPI 设备管理接口来访问 SPI 设备。RT-Thread 中的 SPI 设备管理接口主要包含查找 SPI 设备、传输数据等。注意：在数据传输中，SPI 设备管理接口会调用 rt_mutex_take() 函数，该函数不能在中断服务程序中调用，否则会导致断言错误。

1）查找 SPI 设备

在使用 SPI 设备前需要根据 SPI 设备名称获取 SPI 设备句柄。注册到 RT-Thread 的 SPI 设备名称通常为 spi10、qspi10 等。通过下面的函数可查找 SPI 设备：

```
/**************************************************************************
 * 名称：rt_device_find()
 * 功能：查找 SPI 设备
 * 参数：name 表示 SPI 设备名称
 * 返回：SPI 设备句柄表示成功查找到对应 SPI 设备；RT_NULL 表示没有找到对应的 SPI 设备
 **************************************************************************/
rt_device_t rt_device_find(const char* name);
```

2）数据传输

获取到 SPI 设备句柄就可以使用 SPI 设备管理接口访问 SPI 设备，进而实现数据传输。通过下面的函数可实现数据传输：

```
/**************************************************************************
 * 名称：rt_spi_transfer_message()
 * 功能：数据传输
 * 参数：device 表示 SPI 设备句柄；message 表示指向待传输数据的结构体指针
 * 返回：RT_NULL 表示数据传输；非空指针表示数据传输失败，返回指向剩余未发送数据的结构体
指针
 **************************************************************************/
struct  rt_spi_message  *rt_spi_transfer_message(struct  rt_spi_device  *device,  struct  rt_spi_message
*message);
```

函数 rt_spi_transfer_message() 可以传输一连串的数据，用户可以自定义每个待传输数据的结构体参数，从而方便地控制数据传输方式。struct rt_spi_message 结构体如下：

```
struct rt_spi_message
{
    const void *send_buf;                        //指向发送缓冲区的指针
    void *recv_buf;                              //指向接收缓冲区的指针
    rt_size_t length;                            //待传输数据的长度，单位为 B
    struct rt_spi_message *next;                 //指向继续传输的下一条数据的指针
```

```
        unsigned cs_take    : 1;                        //片选选中
        unsigned cs_release : 1;                        //释放片选
};
```

其中，send_buf 表示指向发送缓冲区的指针，其值为 RT_NULL 时，表示只接收数据，不需要发送数据；recv_buf 表示指向接收缓冲区的指针，其值为 RT_NULL 时，表示只发送数据，不需要保存接收到的数据，将接收到的数据直接丢弃；length 表示待传输数据的长度；next 表示指向继续发送的下一条数据的指针，若只发送一条消息，则 next 的值为 RT_NULL，多个待传输的数据通过 next 指针以单向链表的形式链接在一起；cs_take 为 1 时，表示在传输数据前设置对应的 CS 信号处于有效状态，cs_release 为 1 时，表示在数据传输结束后释放对应的 CS 信号。

注意：当 send_buf 或 recv_buf 不为空时，两者的可用空间都不得小于 length。若使用函数 rt_spi_transfer_message() 传输数据，在传输的第一条消息前需要将 cs_take 设置为 1，可使 CS 信号处于有效状态，在传输最后一条数据后需要将 cs_release 设置为 1，从而释放对应的 CS 信号。

RT-Thread 还提供了以下的函数来控制 SPI 设备发送或者接收数据，这些函数都是对函数 rt_spi_transfer_message() 的再封装。

- ⊃ rt_spi_transfer()：传输一次数据。
- ⊃ rt_spi_send()：发送一次数据。
- ⊃ rt_spi_recv()：接收一次数据。
- ⊃ rt_spi_send_then_send()：连续两次发送数据。
- ⊃ rt_spi_send_then_recv()：先发送数据再接收数据。

3.9.1.4　特殊场景的 SPI 设备管理接口

1）获取 SPI 总线的使用权

在多线程的情况下，同一条 SPI 总线可能会在被不同的线程使用。为了防止丢失正在传输的数据，SPI 从机在开始传输数据前需要先获取 SPI 总线的使用权，获取成功后才能传输数据。通过下面的函数可获取 SPI 总线的使用权：

```
/**********************************************************************************
 * 名称：rt_spi_take_bus()
 * 功能：获取 SPI 总线的使用权
 * 参数：device 表示 SPI 设备句柄
 * 返回：RT_EOK 表示成功；其他错误码表示失败
 **********************************************************************************/
rt_err_t rt_spi_take_bus(struct rt_spi_device *device);
```

2）选中 SPI 从机

SPI 从机在获取 SPI 总线的使用权后，需要将自己的 CS 信号设置为有效。通过下面的函数可设置 CS 信号：

```
/**********************************************************************************
 * 名称：rt_spi_take()
 * 功能：设置 CS 信号
 * 参数：device 表示 SPI 设备句柄
```

```
*****************************************************************************/
rt_err_t rt_spi_take(struct rt_spi_device *device);
```

3）增加一条待传输的数据

在使用 rt_spi_transfer_message()函数传输数据时，待传输的数据是以单向链表的形式链接起来的。通过下面的函数可在单向链表中增加一条新的待传输数据：

```
/*****************************************************************************
 * 名称：rt_spi_message_append()
 * 功能：增加一条待传输的数据
 * 参数：list 表示待传输数据的单向链表；message 表示指向新增消息的指针
 *****************************************************************************/
rt_evoid rt_spi_message_append(struct rt_spi_message *list, struct rt_spi_message *message);
```

4）释放 SPI 从机

SPI 从机传输完数据后，需要释放 CS 信号来释放 SPI 从机。通过下面的函数可释放 CS 信号：

```
/*****************************************************************************
 * 名称：rt_spi_release()
 * 功能：释放 CS 信号
 * 参数：device 表示 SPI 设备句柄
 *****************************************************************************/
rt_err_t rt_spi_release(struct rt_spi_device *device);
```

5）释放 SPI 总线

当 SPI 从机不再使用 SPI 总线时，必须尽快释放 SPI 总线，这样其他 SPI 从机才能使用 SPI 总线传输数据。通过下面的函数可释放 SPI 总线：

```
/*****************************************************************************
 * 名称：rt_spi_release_bus()
 * 功能：释放 SPI 总线
 * 参数：device 表示 SPI 设备句柄
 * 返回：RT_EOK 表示成功
 *****************************************************************************/
rt_err_t rt_spi_release_bus(struct rt_spi_device *device);
```

3.9.2　开发设计与实践

3.9.2.1　硬件设计

本节通过 MobaXterm 串口终端 FinSH 控制台来控制 SPI 设备，帮助读者掌握 SPI 设备管理接口的应用。FinSH 控制台和 ZI-ARMEmbed 的连接如图 3.32 所示。

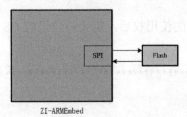

图 3.32　硬件设计

本节的 Flash 芯片使用的是 W25Q64，其硬件连接电路如图 3.33 所示。

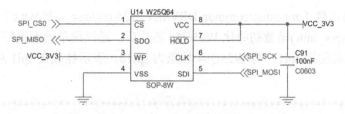

图 3.33　W25Q64 的硬件连接电路

3.9.2.2　软件设计

软件设计流程如图 3.34 所示。

（1）在主函数中调用 SPI 设备的管理接口。

（2）先挂载 SPI 设备，再查找 SPI 设备、配置 SPI 设备，接着发送、读取数据，最后显示读取到的数据。

图 3.34　软件设计流程

3.9.2.3　功能设计与核心代码设计

要控制 SPI 设备，需要使用 RT-Thread 提供的 SPI 设备管理接口。

1）主函数（zonesion/app/main.c）

主函数的主要工作是调用 w25qxx_init()函数，完成应用程序的调用。代码如下：

```
/*****************************************************************************
* 文件：main.c
*****************************************************************************/
#include <rtthread.h>
int main(void)
{
    w25qxx_init();
    rt_uint32_t FLASH_SIZE=8*1024*1024;    //W25Q64 的容量为 8 MB，最大地址是 8 × 1024 × 1024
    w25qxx_readwrite_sample(1,FLASH_SIZE-4096-10);
    return 0;
```

```
    }
```

2）SPI 控制函数（zonesion/common/DRV/dev_w25qxx/dev_w25qxx.c）

在使用 w25qxx_init()函数初始化 W25Q64 芯片后，通过简单的配置即可启动 SPI 总线。这里的 SPI 总线名称为 spi1，将 W25Q64 抽象为 spi10，表示挂载在 spi1 总线上的 W25Q64 是 spi10 设备。

```c
/*******************************************************************************
 * 名称：w25qxx_init()
 * 功能：SPI 设备的相关操作，包括初始化、配置、写入、读取、显示
 * 返回：0 表示成功；-1 表示失败
 ******************************************************************************/
#include "dev_w25qxx.h"
#ifdef RT_USING_SPI
#include <drivers/spi.h>
#include "drv_spi.h"

rt_uint16_t W25QXX_TYPE=W25Q64;                                    //默认是 W25Q64

struct rt_spi_device *w25qxxDev = RT_NULL;
static uint8_t w25qxx_init_ok = 0;
static uint8_t cmd;

//注意 W25Q128 的处理地址不同，这里只适配了 W25Q64
int w25qxx_init(void)
{
    if(w25qxx_init_ok == 1) return 0;
    //挂载在 SPI 总线上，抽象为 spi10 设备
    rt_hw_spi_device_attach("spi1", "spi10", GPIOA, GPIO_PIN_15);

    //查找 SPI 设备
    w25qxxDev = (struct rt_spi_device *)rt_device_find(SPI_DEV_NAME);
    if(w25qxxDev == RT_NULL)
    {
        rt_kprintf("failed to find %s device!\n", SPI_DEV_NAME);
        return -1;
    }
    struct rt_spi_configuration cfg;
    cfg.data_width = 8;                          //SPI 总线的数据宽度
    cfg.mode = RT_SPI_MODE_3 | RT_SPI_MSB;       //SPI 总线的工作模式，支持工作模式 0 和 3
    cfg.max_hz = 50 * 1000 * 1000;               //SPI 总线的最大工作频率
    rt_spi_configure(w25qxxDev, &cfg);           //配置 SPI 总线

    W25QXX_TYPE= W25QXX_ReadID();
    rt_kprintf("read W25Qxx ID is: 0x%04X\n", W25QXX_TYPE);        //读取 SPI 设备 ID
    w25qxx_init_ok = 1;
    rt_thread_delay(100);                        //延时
    return 0;
```

```
}
```

3）自定义 FinSH 控制台命令函数（zonesion/app/finsh_cmd/finsh_ex.c）

自定义 FinSH 控制台命令函数的主要工作是实现 SPI 设备的管理命令，可在 FinSH 控制台中使用这些命令来控制 SPI 设备。代码如下：

```c
/******************************************************************************
* 名称：spiflash()
* 功能：自定义 FinSH 控制台命令
* 参数：argc 表示参数个数；argv 表示指向参数字符串指针的数组指针
* 返回：0 表示成功
******************************************************************************/
int spiflash(int argc, char **argv)
{
    if(!rt_strcmp(argv[1], "write"))
    {
        if(argc < 3)                                    //检测命令是否包含参数
        {
            rt_kprintf("Please enter write <address data>!\n");
            return 0;
        }
        rt_uint32_t address=atoi(argv[2]);
        if(address>FLASH_SIZE || address<0){
            rt_kprintf("address exceeds length!\n");
            return 0;
        }
        char tempBuf[128] = {0};
        snprintf(tempBuf, sizeof(tempBuf), "%s\n", argv[3]);
        W25QXX_Write((rt_uint8_t*)tempBuf,address,strlen(tempBuf));
    }
    else if(!rt_strcmp(argv[1], "read"))
    {
        rt_uint32_t address=atoi(argv[2]);
        rt_uint16_t length=atoi(argv[3]);
        if(address>FLASH_SIZE || address<0){
            rt_kprintf("address exceeds length!\n");
            return 0;
        }
        if(length>1000 || length<=0){
            rt_kprintf("read length is too long!\n");
            return 0;
        }
        char* readdata=rt_malloc(length+1);
        memset(readdata,0,length+1);
        W25QXX_Read((rt_uint8_t*)readdata,address,length);
        rt_kprintf("readdata is: %s \r\n",readdata);
    }
    else
```

```
        rt_kprintf("Please input 'spiflash <write|read>'\n");
    return 0;
}
MSH_CMD_EXPORT(spiflash, spiflash readwrite sample);
```

3.9.3　开发步骤与验证

3.9.3.1　硬件部署

同 1.2.3.1 节。

3.9.3.2　项目部署

（1）将本项目工程文件夹复制到 RT-ThreadStudio\workspace 工作区目录下。

（2）其余同 2.1.3.2 节。

3.9.3.3　验证效果

（1）关闭 RT-Thread Studio，拔掉仿真器，按下 ZI-ARMEmbed 上的电源按键重新上电。

（2）在 MobaXterm 串口终端 FinSH 控制台中输入命令"spiflash"，解析自定义 FinSH 控制台命令后，FinSH 控制台将输出"Please input 'spiflash <write|read>'"。

```
 \ | /
- RT -     Thread Operating System
 / | \      4.1.0 build Oct 31 2022 16:00:34
 2006 - 2022 Copyright by RT-Thread team
read W25Qxx ID is: 0xEF16
msh >spiflash
Please input 'spiflash <write|read>'
```

（3）在 MobaXterm 串口终端 FinSH 控制台中输入命令"spiflash write 4094 comeon123"，解析自定义 FinSH 控制台命令后，以 SPI 设备的 4094 地址为起始地址开始写入数据（数据应为字符串格式）。

```
 \ | /
- RT -     Thread Operating System
 / | \      4.1.0 build Oct 31 2022 16:00:34
 2006 - 2022 Copyright by RT-Thread team
read W25Qxx ID is: 0xEF16
msh >spiflash
Please input 'spiflash <write|read>'
msh >spiflash write 4094 comeon123
msh >
```

（4）在 MobaXterm 串口终端 FinSH 控制台中输入命令"spiflash read 4094 9"，解析自定义 FinSH 控制台命令后，可读取 SPI 设备中的数据（读取格式为起始地址和数据长度），并在 FinSH 控制台中输出"readdata is: comeon123"。

```
 \ | /
- RT -     Thread Operating System
 / | \      4.1.0 build Oct 31 2022 16:00:34
```

3.9.4　小结

本节主要介绍 SPI 设备的基本概念和管理方式。通过本节的学习，读者可掌握 RT-Thread SPI 设备管理接口的应用。

3.10　WATCHDOG 设备驱动应用开发

看门狗（Watchdog）是一种检测和控制机制，旨在检测系统是否正常运行，并在检测到问题时采取预设的纠正措施。这种机制对于确保系统的稳定性和可靠性非常重要。WATCHDOG 设备广泛应用于嵌入式系统、服务器、网络设备等需要高稳定性和可靠性的场景。

本节的要求如下：

- 了解 WATCHDOG 设备的基本概念。
- 掌握 WATCHDOG 设备的管理方式。
- 掌握 RT-Thread WATCHDOG 设备管理接口的应用。

3.10.1　原理分析

3.10.1.1　WATCHDOG 设备简介

硬件看门狗：硬件看门狗是一个定时器，其定时输出连接到电路的复位端。硬件看门狗可以通过定期接收来自系统的信号来检测系统是否正常运行。如果系统停止响应或发生其他问题，则硬件看门狗可执行一些预设的操作，如强制系统重启。

软件看门狗：软件看门狗是通过软件实现的一种看门狗。在应用程序层面，程序员可以设置一个定时器，定期向看门狗发送信号，以表示应用程序在正常运行。如果应用程序出现故障或停滞，软件看门狗可能会采取预设的纠正措施，如重启应用程序或执行其他恢复操作。

当看门狗启动后，定时器开始自动计数，如果定时器在溢出前没有被复位，则定时器溢出会对 CPU 产生一个复位信号使系统重启（俗称被狗咬）。在系统正常运行时，需要在看门狗允许的时间间隔内将看门狗定时器清零（俗称喂狗），不产生复位信号。如果系统不出问题，则程序能够定期喂狗。一旦程序跑飞，没有喂狗，系统将复位。

看门狗的工作原理如下：

（1）定时器设定：看门狗具有内置的定时器，用于周期性计时。

（2）定期喂狗：软件或系统中的应用程序需要定期喂狗，也就是向看门狗发送信号，告诉它系统正常运行。这个信号可以是一个特殊的命令或简单的定时器重置信号。

（3）监控信号：看门狗不断检测收到的喂狗信号，只要系统正常运行，就会定期收到喂狗信号。

（4）异常检测：如果看门狗在预定时间内没有收到喂狗信号，就会认为系统处于异常状态。异常可能是由于系统崩溃、死锁或其他问题引起的。

（5）纠正措施：一旦检测到异常，看门狗将采取一些预设的纠正措施，如强制系统重启、恢复安全状态或者执行其他操作。

（6）系统复位：在很多场合中，看门狗的主要功能之一是强制系统重启，以便系统能够从头开始，并恢复到正常状态。

一般情况下，可以在 RT-Thread 的 idle 回调函数和关键任务中喂狗。

3.10.1.2　访问 WATCHDOG 设备

1）查找 WATCHDOG 设备

应用程序可根据 WATCHDOG 设备名称获取 WATCHDOG 设备句柄，进而操作 WATCHDOG 设备。通过下面的函数可查找 WATCHDOG 设备：

```
/*********************************************************************
 * 名称：rt_device_find()
 * 功能：查找 WATCHDOG 设备
 * 参数：name 表示 WATCHDOG 设备名称
 * 返回：WATCHDOG 设备句柄表示成功查找到对应的 WATCHDOG 设备；RT_NULL 表示没有找
到对应的 WATCHDOG 设备
 *********************************************************************/
rt_device_t rt_device_find(const char* name);
```

2）初始化 WATCHDOG 设备

在使用 WATCHDOG 设备前需要先对其进行初始化。通过下面的函数可初始化 WATCHDOG 设备：

```
/*********************************************************************
 * 名称：rt_device_init()
 * 功能：初始化 WATCHDOG 设备
 * 参数：dev 表示 WATCHDOG 设备句柄
 * 返回：RT_EOK 表示 WATCHDOG 设备初始化成功；-RT_ENOSYS 表示 WATCHDOG 设备初始
化失败；其他错误码表示 WATCHDOG 设备打开失败
 *********************************************************************/
rt_err_t rt_device_init(rt_device_t dev);
```

3）控制 WATCHDOG 设备

应用程序可通过命令控制字来控制 WATCHDOG 设备。通过下面的函数可控制 WATCHDOG 设备：

```
/*********************************************************************
 * 名称：rt_device_control()
 * 功能：控制 WATCHDOG 设备
 * 参数：dev 表示 WATCHDOG 设备句柄；cmd 表示命令控制字；arg 表示命令控制字的参数
```

* 返回：RT_EOK 表示函数执行成功；-RT_ENOSYS 表示执行失败（dev 为空）；其他错误码表示执行失败
**/
rt_err_t rt_device_control(rt_device_t dev, rt_uint8_t cmd, void* arg);

命令控制字可取以下的宏定义：

#define RT_DEVICE_CTRL_WDT_GET_TIMEOUT	(1)	//获取溢出时间
#define RT_DEVICE_CTRL_WDT_SET_TIMEOUT	(2)	//设置溢出时间
#define RT_DEVICE_CTRL_WDT_GET_TIMELEFT	(3)	//获取剩余时间
#define RT_DEVICE_CTRL_WDT_KEEPALIVE	(4)	//喂狗
#define RT_DEVICE_CTRL_WDT_START	(5)	//启动 WATCHDOG 设备
#define RT_DEVICE_CTRL_WDT_STOP	(6)	//停止 WATCHDOG 设备

4）关闭 WATCHDOG 设备

当应用程序完成对 WATCHDOG 设备的操作后，可以关闭 WATCHDOG 设备。通过下面的函数可关闭 WATCHDOG 设备：

/**
* 名称：rt_device_close()
* 功能：关闭 WATCHDOG 设备
* 参数：dev 表示 WATCHDOG 设备句柄
* 返回：RT_EOK 表示关闭 WATCHDOG 设备成功；-RT_ERROR 表示 WATCHDOG 设备失败
（WATCHDOG 设备已经被关闭了）；其他错误码表示关闭 WATCHDOG 设备失败
**/
rt_err_t rt_device_close(rt_device_t dev);

3.10.2　开发设计与实践

3.10.2.1　硬件设计

本节通过 MobaXterm 串口终端 FinSH 控制台来控制 WATCHDOG 设备，帮助读者掌握 WATCHDOG 设备管理接口的应用。FinSH 控制台、LED 和 ZI-ARMEmbed 的连接如图 3.35 所示。

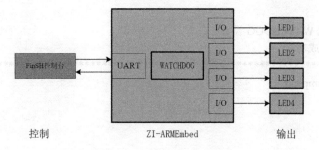

图 3.35　FinSH 控制台、LED 和 ZI-ARMEmbed 的连接

LED 和 ZI-ARMEmbed 的连接同 1.5.2.1 节。

3.10.2.2　软件设计

软件设计流程如图 3.36 所示。

（1）初始化 LED 及 WATCHDOG 设备。

（2）在 MobaXterm 串口终端 FinSH 控制台中输入控制命令来控制 WATCHDOG 设备，从而控制 LED。

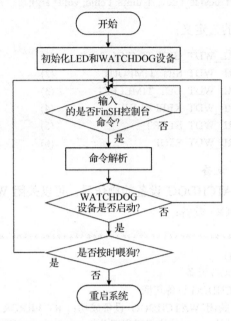

图 3.36　软件设计流程

3.10.2.3　功能设计与核心代码设计

要控制 WATCHDOG 设备，需要使用 RT-Thread 提供的 WATCHDOG 设备管理接口完成对 WATCHDOG 设备的初始化，并编写相应的控制函数。要通过 FinSH 控制台命令对 WATCHDOG 设备进行控制，需要自定义相应的控制命令，并将控制命令添加到 FinSH 控制台命令列表中。

1）主函数（zonesion/app/main.c）

主函数的主要工作是调用 WATCHDOG 设备的初始化函数。代码如下：

```
/*******************************************************************************
* 名称：main()
* 功能：初始化 WATCHDOG 设备
* 返回：0 表示成功
*******************************************************************************/
#include "wdt_sample.h"
int main(void)
{
    wdt_init();
    return 0;
}
```

2）WATCHDOG 设备初始化函数（zonesion/app/wdt_sample.c）

WATCHDOG 设备初始化函数的主要工作是初始化 LED、初始化和配置 WATCHDOG 设备、点亮 LED（用于指示程序执行状态）。代码如下：

```
/**************************************************************************
* 名称：wdt_init()
* 功能：初始化 WATCHDOG 设备
* 返回：0 表示初始化成功；-1 表示初始化失败
**************************************************************************/
#include "wdt_sample.h"
#include "drv_led.h"
rt_device_t wdgDev = RT_NULL;
int wdt_init(void)
{
    led_pin_init();                                                 //LED 初始化
    wdgDev = rt_device_find(WDT_DEV_NAME);                          //查找 WATCHDOG 设备
    if(wdgDev == RT_NULL)
    {
        rt_kprintf("failed to find %s device!\n", WDT_DEV_NAME);
        return -1;
    }
    rt_err_t result = RT_EOK;
    rt_uint32_t timeout = 10;
    result = rt_device_init(wdgDev);                                //初始化 WATCHDOG 设备
    if(result != RT_EOK)
    {
        rt_kprintf("initialize %s failed!\n", WDT_DEV_NAME);
        return RT_ERROR;
    }
    //设置 WATCHDOG 设备的溢出时间
    result = rt_device_control(wdgDev, RT_DEVICE_CTRL_WDT_SET_TIMEOUT, &timeout);
    if(result != RT_EOK)
    {
        rt_kprintf("set %s timeout failed!\n", WDT_DEV_NAME);
        return RT_ERROR;
    }
    else
        rt_kprintf("watchdog timeout period: %d sec\n", timeout);
    rt_thread_mdelay(1000);                                         //延时 1000 ms
    led_ctrl(0x0F);                                                 //点亮 LED
    return 0;
}
```

3）自定义 FinSH 控制台命令函数（zonesion/app/finsh_cmd/finsh_ex.c）

自定义 FinSH 控制台命令函数的主要工作是实现 WATCHDOG 设备的管理命令，可在 FinSH 控制台中使用这些命令来控制 WATCHDOG 设备。代码如下：

```
/**************************************************************************
* 名称：wdtDrv()
* 功能：自定义 WATCHDOG 设备的 FinSH 控制台命令
* 参数：argc 表示参数个数；argv 表示指向参数字符串指针的数组指针
**************************************************************************/
```

```
int wdtDrv(int argc, char **argv)
{
    if(!rt_strcmp(argv[1], "start"))
    {
        //启动 WATCHDOG 设备
        rt_err_t result = rt_device_control(wdgDev, RT_DEVICE_CTRL_WDT_START, RT_NULL);
        if(result != RT_EOK)
        {
            rt_kprintf("watchdog start failed!\n");
        } else {
            rt_kprintf("watchdog start success!\n");
        }
    }
    else if(!rt_strcmp(argv[1], "feed"))
    {
        rt_device_control(wdgDev, RT_DEVICE_CTRL_WDT_KEEPALIVE, RT_NULL);        //喂狗
        rt_kprintf("feed the dog!\n");
    }
    else if(!rt_strcmp(argv[1], "set"))
    {
        if(argc < 3)                                        //检测命令是否包含参数
        {
            rt_kprintf("Please enter the timeout period(unit: S)!\n");
            return 0;
        }
        rt_err_t result = RT_EOK;
        rt_uint32_t timeout = atoi(argv[2]);                //获取参数值
        //设置 WATCHDOG 设备的溢出时间
        result=rt_device_control(wdgDev,RT_DEVICE_CTRL_WDT_SET_TIMEOUT,&timeout);
        if (result != RT_EOK)
        {
            rt_kprintf("set %s timeout failed!\n", WDT_DEV_NAME);
            return RT_ERROR;
        }
    }
    else
        rt_kprintf("Please input 'wdtDrv <feed|set>'\n");
    return 0;
}
MSH_CMD_EXPORT(wdtDrv, watchdog driver sample);//将 wdtDrv 命令添加到 FinSH 控制台命令列表中
```

3.10.3　开发步骤与验证

3.10.3.1　硬件部署

同 1.2.3.1 节。

3.10.3.2　工程调试

（1）将本项目工程文件夹复制到 RT-ThreadStudio\workspace 工作区目录下。

（2）其余同 2.1.3.2 节。

3.10.3.3　验证效果

（1）关闭 RT-Thread Studio，拔掉仿真器，按下 ZI-ARMEmbed 上的电源按键重新上电。

（2）在 MobaXterm 串口终端 FinSH 控制台中输入"wdtDrv start"命令，解析自定义 FinSH 控制台命令后可启动 WATCHDOG 设备，并点亮 4 个 LED（见图 3.37），同时在 FinSH 控制台输出"watchdog start success"。

```
 \ | /
- RT -       Thread Operating System
 / | \       4.1.0 build Oct 25 2022 14:41:37
 2006 - 2022 Copyright by RT-Thread team
watchdog timeout period: 10 sec
msh >wdtDrv start
watchdog start success!
msh >
```

图 3.37　点亮 4 个 LED

（3）如果在 10 s 内没有喂狗操作，WATCHDOG 设备产生的复位信号会使系统重启，在 MobaXterm 串口终端 FinSH 控制台中输出系统重启的信息，4 个 LED 将熄灭 1 s 后重新点亮。

```
 \ | /
- RT -       Thread Operating System
 / | \       4.1.0 build Oct 25 2022 14:41:37
 2006 - 2022 Copyright by RT-Thread team
watchdog timeout period: 10 sec
msh >wdtDrv start
watchdog start success!
msh >
 \ | /
- RT -       Thread Operating System
 / | \       4.1.0 build Oct 25 2022 14:41:37
 2006 - 2022 Copyright by RT-Thread team
watchdog timeout period: 10 sec
msh >
```

（4）如果按时进行喂狗操作，即在 10 s 内在 MobaXterm 串口终端 FinSH 控制台中输入 "wdtDrv feed"，则可在解析自定义 FinSH 控制台命令后，使系统保持正常运行，4 个 LED 不会熄灭。

```
     \|/
    - RT -      Thread Operating System
     /|\        4.1.0 build Oct 25 2022 14:41:37
     2006 - 2022 Copyright by RT-Thread team
    watchdog timeout period: 10 sec
    msh >wdtDrv start
    watchdog start success!
    msh >
     \|/
    - RT -      Thread Operating System
     /|\        4.1.0 build Oct 25 2022 14:41:37
     2006 - 2022 Copyright by RT-Thread team
    watchdog timeout period: 10 sec
    msh >wdtDrv feed
    feed the dog!
    msh >
```

（5）在 MobaXterm 串口终端 FinSH 控制台中输入 "wdtDrv set 20"，可以将喂狗的溢出时间修改为 20 s。再次输入 "wdtDrv start" 可验证超时时间的变化，发现溢出时间由原来的 10 s 变成了 20 s。

```
     \|/
    - RT -      Thread Operating System
     /|\        4.1.0 build Oct 25 2022 14:41:37
     2006 - 2022 Copyright by RT-Thread team
    watchdog timeout period: 10 sec
    msh >wdtDrv start
    watchdog start success!
    msh >
     \|/
    - RT -      Thread Operating System
     /|\        4.1.0 build Oct 25 2022 14:41:37
     2006 - 2022 Copyright by RT-Thread team
    watchdog timeout period: 10 sec
    msh >wdtDrv feed
    feed the dog!
    msh >

    msh >wdtDrv set 20
    msh >
```

3.10.4   小结

本节主要介绍 WATCHDOG 设备的基本概念和管理方式。通过本节的学习，读者可掌握 RT-Thread WATCHDOG 设备管理接口的应用。

3.11 SENSOR 设备驱动应用开发

SENSOR 设备是指传感器设备，这些设备能够感知、测量和检测物理环境中的各种参数和信号，如温度、湿度、光照度、加速度、压力、声音、位置等。SENSOR 设备在各种领域中得到了广泛应用，如工业自动化、医疗设备、消费电子、汽车行业、智能手机、物联网等。

- ⊃ 温度传感器：测量环境温度，广泛用于气象观测、温度控制系统、食品存储和医疗设备。
- ⊃ 湿度传感器：测量环境湿度，广泛用于气象观测、湿度控制系统、农业和食品加工。
- ⊃ 光照度传感器：测量环境光照度，广泛用于自动照明系统、屏幕亮度调节系统、太阳能跟踪系统和安全系统。
- ⊃ 加速度传感器：测量物体的加速度，广泛应用于智能手机、游戏控制器、运动跟踪设备和车辆稳定性控制。
- ⊃ 压力/压强传感器：测量压力/压强变化，用于大气压强测量、高度计、液位控制和医疗设备。
- ⊃ 声音传感器：测量声音和噪声，广泛用于音频录制、噪声检测、语音识别和安全系统。
- ⊃ 位置传感器：测量物体的位置，广泛用于全球定位系统（GPS）、车辆导航、机器人导航和地震检测。
- ⊃ 磁场传感器：测量磁场强度和方向，广泛用于罗盘、导航、地磁测量和地质勘探。
- ⊃ 气体传感器：测量特定气体的浓度，广泛用于空气质量检测、气体泄漏检测和工业过程控制。
- ⊃ 生物传感器：测量生物体内的生理参数，如心率、血糖、血压，广泛用于医疗诊断和检测。

SENSOR 设备通常包括传感器元件和信号处理电路，能够将检测到的物理信号转化为电信号，并将电信号发送到微控制器或计算机系统进行处理。SENSOR 设备对于实现自动化、智能控制、环境检测和数据采集等应用至关重要。SENSOR 设备也是物联网的基础之一，通过 SENSOR 设备，系统可以采集数据并与互联网连接，实现远程监控和智能决策。

本节的要求如下：

- ⊃ 了解 SENSOR 设备的基本概念。
- ⊃ 掌握 SENSOR 设备管理的方式。
- ⊃ 掌握 RT-Thread SENSOR 设备管理接口的应用。

3.11.1 原理分析

3.11.1.1 SENSOR 设备简介

传感器是物联网的重要组成部分。随着物联网的发展，已经有大量的传感器被开发出来供开发者使用，如加速度计、磁力计、陀螺仪、气压计、湿度计等。大部分半导体厂商都生产传感器，虽然增加了用户的选择性，但也加大了应用程序的开发难度。不同的传感器厂商、不同的传感器都需要配套各自的驱动程序才能运转起来，在开发应用程序时需要适配不同的传感器，加大了开发难度。

为了降低应用程序的开发难度，增加传感器驱动程序的可复用性，RT-Thread 设计了 SENSOR 设备。SENSOR 设备的作用是为上层提供统一的操作接口，提高了上层代码的可重用性。SENSOR 设备提供了标准的管理接口，支持多种工作模式和电源模式。

表 3.4 给出了目前已经对接到 RT-Thread SENSOR 设备的传感器。

表 3.4 已经对接到 RT-Thread SENSOR 设备的传感器

传感器厂商	传感器型号	备　注
BOSCH	BMA400	加速度计、计步计
	BMI160	加速度计、陀螺仪
	BMX160	加速度计、陀螺仪、磁力计
	BME280	气压计、湿度计、温度计
Goertek	SPL0601	气压计、温度计
ST	LSM6DSL	加速度计、陀螺仪、计步计
	LSM303AGR	加速度计、磁力计
	HTS221	气压计、气温计
	LPS22HB	气压计、气温计
MiraMEMS	DA270	加速度计
	DF220	压力计
ALPSALPINE	HSHCAL001	湿度计、温度计
MEAS	MS5611	气压计、温度计
InvenSense	MPU6XXX(MPU6050/MPU6000/ICM20608)	加速度计、陀螺仪
ASAIR	AHT10	温度计、湿度计
ROHM	BH1750FVI	环境光照强度
Richtek	RT3020	加速度计

3.11.1.2 访问 SENSOR 设备

1）查找 SENSOR 设备

应用程序可根据 SENSOR 设备名称来获取 SENSOR 设备句柄，进而操作 SENSOR 设备。通过下面的函数可查找 SENSOR 设备：

```
/*********************************************************************************
 * 名称：rt_device_find()
 * 功能：查找 SENSOR 设备
 * 参数：name 表示 SENSOR 设备名称
 * 返回：SENSOR 设备句柄表示查找到对应的 SENSOR 设备；RT_NULL 表示没有查找到对应的 SENSOR
设备
 *********************************************************************************/
rt_device_t rt_device_find(const char* name);
```

2）打开 SENSOR 设备

通过 SENSOR 设备句柄，应用程序可以打开和关闭 SENSOR 设备。在打开 SENSOR 设备时，RT-Thread 会检测 SENSOR 设备是否已经初始化，如果没有被初始化则会调用默认的初始化接口初始化 SENSOR 设备。通过下面的函数可打开 SENSOR 设备：

```
/**************************************************************************
 * 名称：rt_device_open()
 * 功能：打开 SENSOR 设备
 * 参数：dev 表示 SENSOR 设备句柄；oflags 表示 SENSOR 设备模式标志
 * 返回：RT_EOK 表示 SENSOR 设备打开成功；-RT_EBUSY 表示如果 SENSOR 设备在注册时指定
的参数中包括 RT_DEVICE_FLAG_STANDALONE 参数，则该 SENSOR 设备将不允许被重复打开；
-RT_EINVAL 表示不支持的打开参数；其他错误码表示 SENSOR 设备打开失败
 **************************************************************************/
rt_err_t rt_device_open(rt_device_t dev, rt_uint16_t oflags);
```

oflags 参数的可选值如下：

```
#define RT_DEVICE_FLAG_RDONLY      0x001      //轮询模式
#define RT_DEVICE_FLAG_INT_RX      0x100      //中断模式
#define RT_DEVICE_FLAG_FIFO_RX     0x200      //FIFO 模式
```

SENSOR 设备的工作模式有三种：中断模式、轮询模式、FIFO 模式。在使用时，这三种模式只能选其一。若在打开 SENSOR 设备时未将 oflags 参数设置为中断模式或 FIFO 模式，则 SENSOR 设备将使用默认的轮询模式。

在 FIFO 模式下，数据存储在硬件 FIFO 中，系统可一次读取多个数据，这可节省 CPU 的使用资源。FIFO 模式在低功耗模式中非常有用。

3）控制 SENSOR 设备

应用程序可通过命令控制字来控制 SENSOR 设备。通过下面的函数可控制 SENSOR 设备：

```
/**************************************************************************
 * 名称：rt_device_control()
 * 功能：控制 SENSOR 设备
 * 参数：dev 表示 SENSOR 设备句柄；cmd 表示命令控制字；arg 表示命令控制字的参数
 * 返回：RT_EOK 表示执行成功；-RT_ENOSYS 表示执行失败（dev 为空）；其他错误码表示执行失败
 **************************************************************************/
rt_err_t rt_device_control(rt_device_t dev, rt_uint8_t cmd, void* arg);
```

命令控制字可取以下的宏定义：

```
#define   RT_SENSOR_CTRL_GET_ID              //读设备 ID
#define   RT_SENSOR_CTRL_GET_INFO            //获取设备信息
#define   RT_SENSOR_CTRL_SET_RANGE           //设置 SENSOR 设备的测量范围
#define   RT_SENSOR_CTRL_SET_ODR             //设置 SENSOR 设备的数据输出速率
#define   RT_SENSOR_CTRL_SET_POWER           //设置电源模式
#define   RT_SENSOR_CTRL_SELF_TEST           //自检
```

4）设置接收回调函数

通过下面的函数可设置接收回调函数，从而在 SENSOR 设备收到数据时，通知上层应用线程有数据到达。

```
/********************************************************************
 * 名称：rt_device_set_rx_indicate()
 * 功能：设置接收回调函数
 * 参数：dev 表示 SENSOR 设备句柄；rx_ind 表示指向回调函数的指针；dev 表示 SENSOR 设备句柄
（回调函数参数）；size 表示缓冲区数据大小（回调函数参数）
 * 返回：RT_EOK 表示设置成功
 ********************************************************************/
rt_err_t rt_device_set_rx_indicate(rt_device_t dev,rt_err_t (*rx_ind)(rt_device_t dev,rt_size_t size));
```

回调函数由调用者提供，若 SENSOR 设备以中断模式打开，则在 SENSOR 设备接收到
数据时会产生中断，从而调用回调函数，并且把缓冲区中的数据大小放在 size 参数里中，把
SENSOR 设备句柄放在 dev 参数中。

5）读取 SENSOR 设备数据

通过下面的函数可读取 SENSOR 设备数据：

```
/********************************************************************
 * 名称：rt_device_read()
 * 功能：读取 SENSOR 设备数据
 * 参数：dev 表示 SENSOR 设备句柄；pos 表示读取数据偏移量，此参数在 SENSOR 设备未使用；
buffer 表示指向缓冲区的指针，读取到的数据将会保存该指针指向的缓冲区中；size 表示读取的数据大小
 * 返回：大于 0 的值表示成功读取到的数据大小；0 表示需要读取当前线程的 errno 来判断错误状态
 ********************************************************************/
rt_size_t rt_device_read(rt_device_t dev, rt_off_t pos, void* buffer, rt_size_t size);
```

6）关闭 SENSOR 设备

当应用程序完成对 SENSOR 设备操作后，需要关闭 SENSOR 设备。通过下面的函数可
关闭 SENSOR 设备：

```
/********************************************************************
 * 名称：rt_device_close()
 * 功能：关闭 SENSOR 设备
 * 参数：dev 表示 SENSOR 设备句柄
 * 返回：RT_EOK 表示关闭 SENSOR 设备成功；-RT_ERROR 表示 SENSOR 设备已经完全关闭，不
能重复关闭该 SENSOR 设备；其他错误码表示关闭 SENSOR 设备失败
 ********************************************************************/
rt_err_t rt_device_close(rt_device_t dev);
```

注意：关闭 SENSOR 设备的函数和打开 SENSOR 设备的函数需要配对使用，打开一次
SENSOR 设备就要关闭一次 SENSOR 设备，这样 SENSOR 设备才会被完全关闭，否则
SENSOR 设备将处于未关闭状态。

3.11.2　开发设计与实践

3.11.2.1　硬件设计

本节通过 MobaXterm 串口终端 FinSH 控制台来控制 SENSOR 设备（以光照度传感器为
例），帮助读者掌握 SENSOR 设备管理接口的应用。FinSH 控制台、光照度传感器和
ZI-ARMEmbed 的连接如图 3.38 所示。

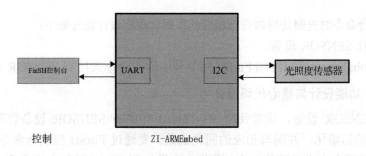

图 3.38　FinSH 控制台、光照度传感器和 ZI-ARMEmbed 的连接

本节使用的光照度传感器是 BH1750FVI-TR，其硬件连接电路如图 3.39 所示。图中，BH1750FVI-TR 的 SCL 引脚与 STM32F407 的 PA1 引脚相连，SDA 引脚与 STM32F407 的 PA0 引脚相连。

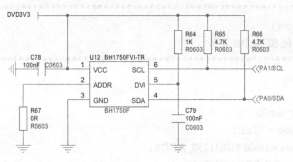

图 3.39　BH1750FVI-TR 的硬件连接电路

3.11.2.2　软件设计

软件设计流程如图 3.40 所示。

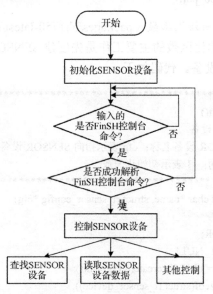

图 3.40　软件设计流程

掌握硬件设计之后，再来分析软件设计。首先进行 SENSOR 初始化工作，包括配置光照传感器、实现相应的操作函数、注册 SENSOR 设备到系统中，再注册 MobaXterm 串口终

端 FinSH 控制台命令对光照传感器进行相应的控制。程序设计流程如下：

（1）初始化 SENSOR 设备。

（2）在 MobaXterm 串口终端 FinSH 控制台中输入命令来控制 SENSOR 设备。

3.11.2.3　功能设计与核心代码设计

要控制 SENSOR 设备，需要使用 RT-Thread 提供的 SENSOR 设备管理接口，完成对 SENSOR 设备的初始化，并编写相应的回调函数。要通过 FinSH 控制台命令对 SENSOR 设备进行控制，需要自定义相应的命令，并将命令添加到 FinSH 控制台命令列表中。这里以光照度传感器为例进行说明。

1）自动初始化 SENSOR 设备函数（packages/bh1750-latest/README.md）

自动初始化 SENSOR 设备函数的主要工作是自动初始化 SENSOR 设备。代码如下。

```
/*****************************************************************************
 * 名称：bh1750_port()
 * 功能：自动初始化 SENSOR 设备
 *****************************************************************************/
int bh1750_port(void)
{
    struct rt_sensor_config cfg;
    cfg.intf.dev_name = "i2c1";
    cfg.intf.user_data = (void *)BH1750_ADDR;
    cfg.irq_pin.pin = RT_PIN_NONE;
    rt_hw_bh1750_init("bh1750", &cfg);
    return 0;
}
INIT_APP_EXPORT(bh1750_port);                          //自动初始化
```

2）SENSOR 设备函数初始化函数（packages/bh1750-latest/sensor_rohm_bh1750.c）

SENSOR 设备函数初始化函数的主要工作是先创建 SENSOR 设备，然后配置 SENSOR 设备，最后注册 SENSOR 设备。代码如下：

```
/*****************************************************************************
 * 名称：rt_hw_bh1750_init()
 * 功能：初始化 SENSOR 设备
 * 参数：name 表示 SENSOR 设备名称；cfg 表示指向 SENSOR 设备配置参数结构体的指针
 * 返回：0 表示初始化成功；-1 表示初始化失败
 *****************************************************************************/
int rt_hw_bh1750_init(const char *name, struct rt_sensor_config *cfg)
{
    int result = -RT_ERROR;
    rt_sensor_t sensor = RT_NULL;
    bh1750_device_t hdev = bh1750_create(&cfg->intf);
    sensor = rt_calloc(1, sizeof(struct rt_sensor_device));
    if (RT_NULL == sensor)
    {
        LOG_E("calloc failed");
        return -RT_ERROR;
```

```
    }
    sensor->info.type = RT_SENSOR_CLASS_LIGHT;
    sensor->info.vendor = RT_SENSOR_VENDOR_UNKNOWN;
    sensor->info.model = "bh1750_light";
    sensor->info.unit = RT_SENSOR_UNIT_LUX;
    sensor->info.intf_type = RT_SENSOR_INTF_I2C;
    sensor->info.range_max = 65535;
    sensor->info.range_min = 1;
    sensor->info.period_min = 120;
    rt_memcpy(&sensor->config, cfg, sizeof(struct rt_sensor_config));
    sensor->ops = &sensor_ops;
    result = rt_hw_sensor_register(sensor, name, RT_DEVICE_FLAG_RDWR, hdev);
    if (result != RT_EOK)
    {
        LOG_E("device register err code: %d", result);
        rt_free(sensor);
        return -RT_ERROR;
    }
    else
    {
        LOG_I("light sensor init success");
        return RT_EOK;
    }
}
```

其中，sensor->ops 指向的结构体包含两个函数指针，分别指向 SENSOR 设备的返回数据和控制函数。

```
static struct rt_sensor_ops sensor_ops =
{
    bh1750_fetch_data,
    bh1750_control
};
```

3）读取 SENSOR 设备数据函数（packages/bh1750-latest/sensor_rohm_bh1750.c）

读取 SENSOR 设备数据函数的主要工作是读取 SENSOR 设备采集到的数据。通过该函数的代码可看到该函数的执行流程，对于光照度传感器而言，该函数依次调用了 bh1750_read_light()、bh1750_read_regs()、rt_i2c_transfer() 和 i2c_bit_xfer() 等函数。

```
/*****************************************************************************
* 名称：bh1750_fetch_data()
* 功能：读取 SENSOR 设备数据
* 参数：sensor 表示 SENSOR 设备句柄；buf 表示指向数据缓冲区的指针；len 表示数据长度
* 返回：1 表示读取数据成功
*****************************************************************************/
static rt_size_t bh1750_fetch_data(struct rt_sensor_device *sensor, void *buf, rt_size_t len)
{
    bh1750_device_t hdev = sensor->parent.user_data;
    struct rt_sensor_data *data = (struct rt_sensor_data *)buf;
```

```
        if (sensor->info.type == RT_SENSOR_CLASS_LIGHT)
        {
            float light_value;
            light_value = bh1750_read_light(hdev);
            data->type = RT_SENSOR_CLASS_LIGHT;
            data->data.light = (rt_int32_t)(light_value * 10);
            data->timestamp = rt_sensor_get_ts();
        }
        return 1;
}
```

4）自定义 FinSH 控制台命令函数（zonesion/app/finsh_cmd/finsh_ex.c）

自定义 FinSH 控制台命令函数的主要工作是实现 SENSOR 设备的管理命令，如包括注册 SENSOR 设备、显示 SENSOR 设备信息、读取 SENSOR 设备数据等，可在 FinSH 控制台中使用这些命令来控制 SENSOR 设备。代码如下：

```
/*****************************************************************************
 * 名称：sensor()
 * 功能：自定义 FinSH 控制台命令
 * 参数：argc 表示参数个数；argv 表示指向参数字符串指针的数组指针
 *****************************************************************************/
static void sensor(int argc, char **argv)
{
    static rt_device_t dev = RT_NULL;
    struct rt_sensor_data data;
    rt_size_t res, i;
    if (argc < 2)           //If the number of arguments less than 2
    {
        ... ...                                             //不合法命令的提示
    }
    else if (!strcmp(argv[1], "info"))                      //显示 SENSOR 设备信息
    {
        ... ...
    }
    else if (!strcmp(argv[1], "read"))                      //读取 SENSOR 设备数据
    {
        uint16_t num = 5;
        if (dev == RT_NULL)
        {
            LOG_W("Please probe sensor device first!");
            return ;
        }
        if (argc == 3)
        {
            num = atoi(argv[2]);
        }
        for (i = 0; i < num; i++)
        {
```

```
                    res = rt_device_read(dev, 0, &data, 1);
                    if (res != 1)
                    {
                        LOG_E("read data failed!size is %d", res);
                    }
                    else
                    {
                        sensor_show_data(i, (rt_sensor_t)dev, &data);
                    }
                    rt_thread_mdelay(100);
                }
            }
            else if (argc == 3)                                    //注册 SENSOR 设备
            {
                if (!strcmp(argv[1], "probe"))
                {
                    rt_uint8_t reg = 0xFF;
                    if (dev)
                    {
                        rt_device_close(dev);
                    }
                    dev = rt_device_find(argv[2]);
                    if (dev == RT_NULL)
                    {
                        LOG_E("Can't find device:%s", argv[1]);
                        return;
                    }
                    if (rt_device_open(dev, RT_DEVICE_FLAG_RDWR) != RT_EOK)
                    {
                        LOG_E("open device failed!");
                        return;
                    }
                    rt_device_control(dev, RT_SENSOR_CTRL_GET_ID, &reg);
                    LOG_I("device id: 0x%x!", reg);
                }
... ...
        }
MSH_CMD_EXPORT(sensor, sensor test function); //将 FinSH 控制台命令添加到 FinSH 控制台命令列表中
```

3.11.3 开发步骤与验证

3.11.3.1 硬件部署

同 1.2.3.1 节。

3.11.3.2 工程调试

（1）将本项目工程文件夹复制到 RT-ThreadStudio\workspace 工作区目录下。

（2）其余同 2.1.3.2 节。

3.11.3.3　验证效果

（1）关闭 RT-Thread Studio，拔掉仿真器，按下 ZI-ARMEmbed 上的电源按键重新上电。

（2）在 MobaXterm 串口终端 FinSH 控制台中输入"sensor probe li_bh1750"，解析自定义 FinSH 控制台命令后，可检测 SENSOR 设备。如果系统中包含该 SENSOR 设备，就会在 FinSH 控制台中输出相应的 ID，并输出"device id:0xff！"。

```
 \|/
- RT -      Thread Operating System
 /|\        4.1.0 build Oct 25 2022 15:47:39
 2006 - 2022 Copyright by RT-Thread team
[I/sensor] rt_sensor[li_bh1750] init success
[I/sensor.rohm.bh1750] light sensor init success
msh >sensor probe li_bh1750
[I/sensor.cmd] device id: 0xff!
msh >
```

（3）在 MobaXterm 串口终端 FinSH 控制台中输入"sensor info"，解析自定义 FinSH 控制台命令后，可获取 SENSOR 设备信息，并在 FinSH 控制台中输出 SENSOR 设备的详细配置信息。

```
 \|/
- RT -      Thread Operating System
 /|\        4.1.0 build Oct 25 2022 15:47:39
 2006 - 2022 Copyright by RT-Thread team
[I/sensor] rt_sensor[li_bh1750] init success
[I/sensor.rohm.bh1750] light sensor init success
msh >sensor probe li_bh1750
[I/sensor.cmd] device id: 0xff!
msh >sensor info
vendor :unknown vendor
model :bh1750_light
unit :lux
range_max :65535
range_min :1
period_min:120ms
fifo_max :0
msh >
```

（4）在 MobaXterm 串口终端 FinSH 控制台中输入"sensor read"，解析自定义 FinSH 控制台命令后，可读取 SENSOR 设备的数据。该命令可以在 FinSH 控制台中连续 5 次输出读取到的 SENSOR 设备数据。

```
 \|/
- RT -      Thread Operating System
 /|\        4.1.0 build Oct 25 2022 15:47:39
 2006 - 2022 Copyright by RT-Thread team
```

```
[I/sensor] rt_sensor[li_bh1750] init success
[I/sensor.rohm.bh1750] light sensor init success
msh >sensor probe li_bh1750
[I/sensor.cmd] device id: 0xff!
msh >sensor info
vendor :unknown vendor
model :bh1750_light
unit :lux
range_max :65535
range_min :1
period_min:120ms
fifo_max :0
msh >sensor read
[I/sensor.cmd] num:   0, light: 1691 lux, timestamp:37499
[I/sensor.cmd] num:   1, light: 1691 lux, timestamp:37805
[I/sensor.cmd] num:   2, light: 1691 lux, timestamp:38111
[I/sensor.cmd] num:   3, light: 1691 lux, timestamp:38417
[I/sensor.cmd] num:   4, light: 1691 lux, timestamp:38723
msh >
```

3.11.4　小结

本节主要介绍 SENSOR 设备的基本概念和管理方式。通过本节的学习，读者可掌握 RT-Thread SENSOR 设备管理接口的应用。

第4章
RT-Thread 文件系统开发技术

4.1 挂载管理应用开发

在操作系统中，挂载管理是指对文件系统与特定目录关联过程的管理。挂载（Mounting）是指将文件系统添加到目录的操作，使用户可以在特定的目录下访问文件系统。

本节的要求如下：

- ➲ 了解虚拟文件系统的基本原理与概念。
- ➲ 了解虚拟文件系统的初始化流程。
- ➲ 掌握挂载文件系统组件 DFS 的管理方式。
- ➲ 掌握 RT-Thread 挂载管理接口的应用。

4.1.1 原理分析

4.1.1.1 虚拟文件系统

在引入挂载管理概念之前，需要先了解虚拟文件系统的概念。设备虚拟文件系统（DFS）是 RT-Thread 提供的虚拟文件系统组件。DFS 采用与 UNIX 类似的文件、文件夹风格。

DFS 的层次架构如图 4.1 所示，主要包括 POSIX 接口层、虚拟文件系统层和设备抽象层。

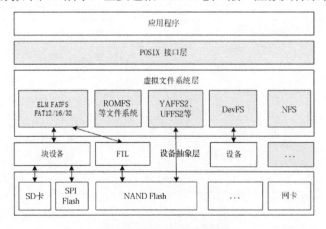

图 4.1　DFS 的层次架构

（1）POSIX 接口层。POSIX 接口是可移植操作系统接口。POSIX 接口标准定义了操作系统应该为应用程序提供的接口标准，旨在获得代码级的软件可移植性。RT-Thread 支持

POSIX 接口，可以很方便地将 Linux、UNIX 的应用程序移植到 RT-Thread 上。

（2）虚拟文件系统层。虚拟文件系统层可以将具体的文件系统注册到 DFS 中，如 FATFS、ROMFS、DevFS 等。

（3）设备抽象层。设备抽象层可以将物理设备（如 SD 卡、SPI Flash、NAND Flash）抽象成文件系统能够访问的设备，如将 FAT 文件系统抽象成块设备。文件系统是独立于存储设备驱动程序实现的，因此把底层存储设备的驱动程序和文件系统对接起来后，才可以正确地使用文件系统。

DFS 的主要功能包括：为应用程序提供了统一的 POSIX 接口和目录操作接口；支持多种类型的文件系统（如 FATFS、ROMFS、DevFS 等），并提供了普通文件、设备文件、网络文件等描述符的管理接口；支持多种类型的存储设备（如 SD 卡、SPI Flash、Nand Flash 等）。

4.1.1.2　挂载管理

文件系统的初始化过程一般是：初始化 DFS→初始化具体的文件系统→在存储设备上创建块设备→格式化块设备→将块设备挂载到 DFS 目录→当不再使用文件系统时卸载相应的块设备。

（1）初始化 DFS。DFS 的初始化是由 dfs_init()函数完成的，该函数会初始化 DFS 所需的相关资源，创建一些关键的数据结构。有了这些数据结构，DFS 便能在系统中找到特定的文件系统（这里以 ELM FATFS 为例进行说明），并获得对特定存储设备内文件的操作方法。如果开启了自动初始化（默认是开启的），则系统将自动调用 dfs_init()函数。

（2）注册文件系统。文件系统的注册过程如图 4.2 所示。

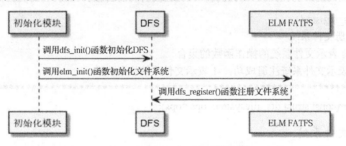

图 4.2　文件系统的注册过程

（3）将存储设备注册为块设备。由于只有块设备才可以挂载到 DFS 目录上，因此需要在存储设备上创建所需的块设备。如果存储设备是 SPI Flash，则可以使用串行 Flash 通用驱动库的 SFUD 模块，该模块提供了各种 SPI Flash 的驱动程序，并将 SPI Flash 抽象成块设备。将存储设备注册为块设备的过程如图 4.3 所示。

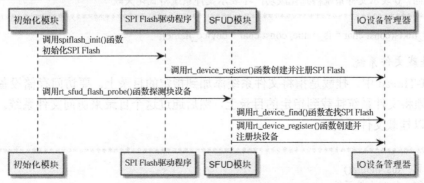

图 4.3　将存储设备注册为块设备的过程

（4）格式化文件系统。在注册块设备后，还需要在块设备上创建指定的文件系统，也就是格式化文件系统。格式化文件系统的过程如图 4.4 所示。

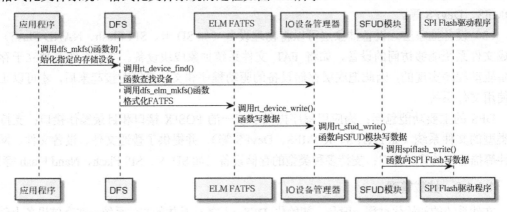

图 4.4　格式化过程

4.1.1.3　挂载管理方式

1）注册文件系统

在 DFS 初始化之后，还需要初始化具体的文件系统，也就是将具体的文件系统注册到 DFS 中。通过下面的函数可注册具体的文件系统：

```
/******************************************************************************
 * 名称：dfs_register()
 * 功能：注册文件系统
 * 参数：ops 表示文件系统的操作函数的集合
 * 返回：0 表示文件系统注册成功；-1 表示文件系统注册失败
 ******************************************************************************/
int dfs_register(const struct dfs_filesystem_ops *ops);
```

2）格式化文件系统

通过下面的函数可格式化文件系统：

```
/******************************************************************************
 * 名称：dfs_mkfs()
 * 功能：格式化文件系统
 * 参数：fs_name 表示文件系统类型；device_name 表示块设备名称
 * 返回：0 表示文件系统格式化成功；-1 表示文件系统格式化失败
 ******************************************************************************/
int dfs_mkfs(const char * fs_name, const char * device_name);
```

3）挂载文件系统

在 RT-Thread 中，挂载是指将文件系统添加到指定的目录上。要访问存储设备中的文件系统，必须将文件系统挂载到指定的目录上，然后通过这个目录来访问文件系统。通过下面的函数可以挂载文件系统：

```
/******************************************************************************
 * 名称：dfs_mount()
 * 功能：挂载文件系统
```

```
    * 参数：device_name 表示已经格式化的块设备名称；path 表示挂载目录，即挂载点；filesystemtype
表示挂载的文件系统类型； rwflag 表示读写标志位；data 表示特定文件系统的私有数据
    * 返回：0 表示文件系统挂载成功；-1 表示文件系统挂载失败
    *************************************************************************/
    int dfs_mount(const char *device_name, const char *path, const char *filesystemtype, unsigned long rwflag,
            const void *data);
```

4）卸载文件系统

当不再使用某个文件系统时，可以卸载该文件系统。通过下面的函数可卸载文件系统：

```
/*************************************************************************
    * 名称：dfs_unmount()
    * 功能：卸载文件系统
    * 参数：specialfile 表示卸载目录
    * 返回：0 表示卸载文件系统成功；-1 表示卸载文件系统失败
    *************************************************************************/
    int dfs_unmount(const char *specialfile);
```

4.1.2　开发设计与实践

4.1.2.1　硬件设计

本节的硬件设计同 3.2.2.1 节。本节通过 MobaXterm 串口终端 FinSH 控制台来控制虚拟文件系统，有助于读者了解文件系统的控制原理、掌握挂载管理的应用。

4.1.2.2　软件设计

软件设计流程如图 4.5 所示。

（1）初始化挂载管理，包括初始化 DFS、文件系统，将存储设备注册为块设备。

（2）在 MobaXterm 串口终端 FinSH 控制台中输入"dfsHandle mkfs"命令，系统调用 dfs_mkfs()函数对文件系统进行格式化。

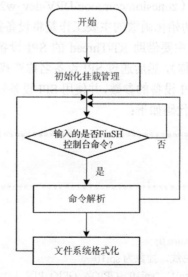

图 4.5　软件设计流程

4.1.2.3　功能设计与核心代码设计

要实现挂载管理，需要使用 RT-Thread 提供的挂载管理接口。本节使用 FinSH 控制台命令实现文件系统的挂载管理，需要自定义 FinSH 控制台命令，并将已实现的命令添加到 FinSH 控制台命令列表中，这样就可以在 MobaXterm 串口终端 FinSH 控制台中使用自定义的命令了。

1）主函数（zonesion\app\main.c）

主函数的主要工作是调用 dfs_init()函数（DFS 初始化函数）、w25qxx_init()函数（Flash 初始化函数）、w25qxx_sfud_init()函数（串行闪存通用驱动初始化函数）。代码如下：

```
/************************************************************************
 * 名称：main()
 ************************************************************************/
#include <rtthread.h>
#include "dfs.h"
#include "dfs_fs.h"
#include "dev_w25qxx.h"

int main(void)
{
    //初始化 DFS，此处只为演示文件挂载流程，实际上在系统初始化时已经调用过此函数了
    dfs_init();
    /*将 SPIFlash 注册为块设备，此处只为演示文件挂载流程，实际上在系统初始化时已经调用过下
面两个函数了*/
    w25qxx_init();
    w25qxx_sfud_init();

    return 0;
}
```

2）W25QXX 初始化函数（zonesion\common\DRV\dev_w25qxx\dev_w25qxx.c）

W25QXX（Flash 芯片）初始化函数的主要工作是将设备挂载到 SPI 总线、查找 SPI 设备和配置 SPI 总线等。该函数主要借助 RT-Thread 的 SPI 设备管理接口，首先将 W25QXX 芯片抽象为 spi10（SPI 设备名称），然后通过 SPI 设备名称查找相应的 SPI 设备指针（句柄），最后以这个指针为基础配置 SPI 设备的参数，并使用 SPI 设备管理接口读取 W25QXX 的 ID，测试 SPI 这些通信是否正常。代码如下：

```
/************************************************************************
 * 名称：w25qxx_init()
 * 功能：W25QXX 初始化
 ************************************************************************/
int w25qxx_init(void)
{
    if(w25qxx_init_ok == 1) return 0;
    //将 SPI 设备挂载到 SPI 总线，抽象为 spi10
    rt_hw_spi_device_attach("spi1", "spi10", GPIOA, GPIO_PIN_15);
    //查找 SPI 设备
```

```
        w25qxxDev = (struct rt_spi_device *)rt_device_find(SPI_DEV_NAME);
        if(w25qxxDev == RT_NULL)
        {
            rt_kprintf("failed to find %s device!\n", SPI_DEV_NAME);
            return -1;
        }
        rt_uint8_t w25qxx_read_id = W25QXX_DeviceID;
        rt_uint8_t id[5] = {0};
        struct rt_spi_configuration cfg;
        cfg.data_width = 8;                                    //SPI 总线的数据宽度
        cfg.mode = RT_SPI_MODE_0 | RT_SPI_MSB;                 //SPI 总线的工作模式
        cfg.max_hz = 50 * 1000 * 1000;                        //SPI 总线的最大工作频率
        rt_spi_configure(w25qxxDev, &cfg);                    //配置 SPI 设备

        //采用先发后读的方式，发送 SPI 设备 ID 后读取 5 B 的数据
        rt_spi_send_then_recv(w25qxxDev, &w25qxx_read_id, 1, id, 5);
        rt_kprintf("read W25Qxx ID is: 0x%02X%02X\n", id[3], id[4]);    //显示读取到的 SPI 设备 ID
        w25qxx_init_ok = 1;
        return 0;
    }
    INIT_COMPONENT_EXPORT(w25qxx_init);
```

3）注册块设备函数和自动挂载列表（zonesion\common\DRV\dev_w25qxx）

首先，通过 w25qxx_sfud_init()函数将 SPI 设备（设备名称为 spi10）注册为块设备（块设备名称为 W25QXX）；然后，系统会自动索引 dfs_mount_tbl mount_table 列表，根据这个列表中的信息挂载指定的块设备；最后，使用 list_device 命令可以查看到注册的块设备。代码如下：

```
/**************************************************************************************
 * 名称：int w25qxx_sfud_init(void)
 * 功能：初始化 W25QXX 块设备
 * 参数：无参数
 * 返回：0 表示参数成功；其他值表示参数失败
 **************************************************************************************/
//自动挂载列表
const struct dfs_mount_tbl mount_table[] =
{
    {"W25QXX", "/", "elm", 0, 0},                 //块设备驱动程序名称、目录、文件系统
    {0}
};
int w25qxx_sfud_init(void)
{
    if(w25qxx_sfud_init_ok == 1) return 0;
    /*将 SPI 设备（设备名称为 spi10）注册为块设备（块设备名称为 W25QXX），通过 SFUD 模块探
测 SPI 设备*/
    if(RT_NULL == rt_sfud_flash_probe("W25QXX", "spi10"))
    {
        return RT_ERROR;
```

```
    }
    w25qxx_sfud_init_ok = 1;
    return RT_EOK;
}
INIT_COMPONENT_EXPORT(w25qxx_sfud_init);
```

4）自定义 FinSH 控制台命令函数（zonesion\app\finsh_cmd\finsh_ex.c）

自定义 FinSH 控制台命令函数的主要工作是实现挂载管理命令，可在 FinSH 控制台中使用这些命令来实现挂载管理。代码如下：

```
/****************************************************************************
* 名称：dfsHandle (int argc, char **argv)
* 功能：自定义 FinSH 控制台命令
* 参数：argc 表示参数个数；argv 表示指向参数字符串指针的数组指针
****************************************************************************/
#include <rtthread.h>
#include "dfs_fs.h"
int dfsHandle(int argc, char **argv)
{
    rt_err_t result = RT_EOK;
    if(!rt_strcmp(argv[1], "mkfs"))
    {
        result = dfs_mkfs("elm", "W25Q64");            //格式化文件系统
        if(result >= 0)
            rt_kprintf("mkfs elm succeed!\n");
        else
            rt_kprintf("mkfs elm fail!\n");
    }
    else
        rt_kprintf("Please input 'dfsHandle <mkfs>'\n");
    return 0;
}
MSH_CMD_EXPORT(dfsHandle, dfs handle);
```

4.1.3　开发步骤与验证

4.1.3.1　硬件部署

同 1.2.3.1 节。

4.1.3.2　工程调试

（1）将本项目工程文件夹复制到 RT-ThreadStudio\workspace 工作区目录下。

（2）其余同 2.1.3.2 节。

4.1.3.3　验证效果

（1）关闭 RT-Thread Studio，拔掉仿真器，按下 ZI-ARMEmbed 上的电源按键重新上电。

（2）ZI-ARMEmbed 重启后，在 MobaXterm 串口终端 FinSH 控制台输出"read W25Qxx

ID is: 0x16EF"（文件系统被挂载到的设备）、"Find a Winbond flash chip. Size is 8388608 bytes."（找到 Winbond 闪存芯片并显示大小）、"W25Q64 flash device is initialize success."（W25QXX 芯片初始化成功）、"Probe SPI flash W25Q64 by SPI device spi10 success."（SPI 设备 spi10 成功探测到 SPI 闪存 W25QXX）。

```
    \ | /
  - RT -      Thread Operating System
   / | \       4.1.0 build Nov   1 2022 15:01:35
  2006 - 2022 Copyright by RT-Thread team
 read W25Qxx ID is: 0x16EF
 [I/SFUD] Find a Winbond flash chip. Size is 8388608 bytes.
 [I/SFUD] W25QXX flash device is initialize success.
 [I/SFUD] Probe SPI flash W25QXX by SPI device spi10 success.
```

（3）在 MobaXterm 串口终端 FinSH 控制台中输入"dfsHandle mkfs"，解析自定义 FinSH 控制台命令后，可格式化文件系统、创建文件系统，并在 FinSH 控制台显示成功格式化文件系统的信息。

```
    \ | /
  - RT -      Thread Operating System
   / | \       4.1.0 build Nov   1 2022 15:01:35
  2006 - 2022 Copyright by RT-Thread team
 read W25Qxx ID is: 0x16EF
 [I/SFUD] Find a Winbond flash chip. Size is 8388608 bytes.
 [I/SFUD] W25QXX flash device is initialize success.
 [I/SFUD] Probe SPI flash W25QXX by SPI device spi10 success.
 dfs already init.
 msh />dfsHandle mkfs
 mkfs elm succeed!
 msh />
```

4.1.4　小结

本节主要介绍文件挂载的基本概念和管理方式。通过本节的学习，读者可掌握 RT-Thread 挂载管理接口的应用。

4.2 文件管理应用开发

文件管理是操作系统的核心功能之一，涉及文件的创建、读取、写入、删除、组织与维护。文件是信息的逻辑单元，而文件系统是在存储设备上组织文件的一种结构。文件系统负责管理文件的存储、检索和组织。

本节的要求如下：

- ○ 了解文件管理的常见操作。
- ○ 掌握文件管理的操作方式。

○ 掌握 RT-Thread 文件管理接口的应用。

4.2.1　原理分析

4.2.1.1　文件管理的常用操作

文件管理是操作系统的一个关键组成部分，它涉及管理计算机系统中的文件和目录，以便用户和应用程序可以创建、访问、修改和删除文件。文件管理的任务包括文件的组织、保护、备份、恢复，以及提供对文件的访问接口。文件管理的常用操作如下：

（1）文件系统组织：文件系统可以看成文件和目录的组织结构，用于在物理存储设备上存储和管理文件。文件管理负责分配空间、维护文件的元数据（如文件名、大小、创建时间、修改时间等），以及记录文件的物理位置。

（2）文件和目录操作：文件管理允许应用程序执行各种文件和目录操作，包括文件的创建、删除、复制、移动等。

（3）文件权限和保护：操作系统允许管理员和文件所有者设置文件的权限，以控制哪些用户或应用程序可以访问文件。文件权限通常包括读、写、执行和特权操作等。

（4）文件备份和还原：文件管理通常支持文件的备份和还原，以防止数据丢失。

（5）文件系统完整性：文件管理可确保文件系统的完整性，以防止文件损坏或丢失，包括文件系统检查和修复工具。

（6）文件索引：文件管理通常会维护文件的索引，以便快速查找文件。索引可以是文件名、目录结构、哈希值等。

（7）文件访问方法：文件管理提供了多种文件访问方法，包括顺序访问、随机访问、直接访问等，以满足不同应用的需求。

4.2.1.2　RT-Thread 的文件管理

RT-Thread 提供了一系列基本的文件操作，包括文件的创建、打开、读取、写入、关闭和删除等。本节主要介绍与文件操作相关的函数，文件操作一般要基于文件描述符 fd 进行。RT-Thread 的常用文件操作如图 4.7 所示。

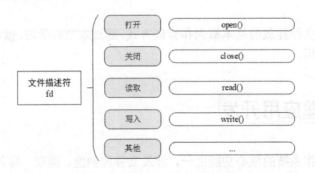

图 4.6　RT-Thread 的常用文件操作

4.2.1.3　RT-Thread 的文件管理方式

1）打开文件

通过下面的函数可打开或创建一个文件：

```
/*****************************************************************************
 * 名称：open()
 * 功能：打开文件
 * 参数：file 表示打开或创建的文件名；flags 表示指定的文件打开方式
 * 返回：文件描述符表示文件打开或创建成功；-1 表示文件打开或创建失败
 *****************************************************************************/
int open(const char *file, int flags, ...);
```

2）关闭文件

当不再需要使用文件后，可通过 close()函数关闭该文件，该函数会将数据写回存储器并释放该文件所占用的资源。

```
/*****************************************************************************
 * 名称：close()
 * 功能：关闭文件
 * 参数：fd 表示文件描述符
 * 返回：0 表示文件关闭成功；-1 表示文件关闭失败
 *****************************************************************************/
int close(int fd);
```

3）读写数据

通过下面的函数可读取文件中的数据：

```
/*****************************************************************************
 * 名称：read()
 * 功能：读取文件
 * 参数：fd 表示文件描述符；buf 表示指向缓冲区的指针；len 表示读取文件的数据字节数
 * 返回：大于 0 的整数表示实际读取到的数据字节数；0 表示读取数据已到达文件结尾或者无可读取
的数据；-1 表示读取数据出错，错误代码可查看当前线程的 errno
 *****************************************************************************/
int read(int fd, void *buf, size_t len);
```

4）写入数据

通过下面的函数可向指定的文件中写入数据：

```
/*****************************************************************************
 * 名称：write()
 * 功能：向文件中写入数据
 * 参数：fd 表示文件描述符；buf 表示指向缓冲区的指针；len 表示写入文件的数据字节数
 * 返回：大于 0 的整数表示实际写入的数据字节数；-1 表示写入出错，错误代码可查看当前线程的
errno
 *****************************************************************************/
int write(int fd, const void *buf, size_t len);
```

4.2.2　开发设计与实践

4.2.2.1　硬件设计

本节的硬件设计同 3.2.2.1 节。本节通过 MobaXterm 串口终端 FinSH 控制台来控制虚拟文件系统，有助于读者掌握文件管理的应用。

4.2.2.2 软件设计

软件设计流程如图 4.7 所示。

（1）初始化文件系统。

（2）在 MobaXterm 串口终端 FinSH 控制台中输入"dfsHandle helps"，FinSH 控制台输出"Please input 'dfsHandle <help|open|close|read|write|unlink>'"。

（3）在 MobaXterm 串口终端 FinSH 控制台中输入"dfsHandle open"，系统会打开或创建一个文件。

（4）在 MobaXterm 串口终端 FinSH 控制台中输入"dfsHandle write"命令时，系统会向文件中写入数据并关闭文件。

（5）在 MobaXterm 串口终端 FinSH 控制台中输入"dfsHandle read"命令时，系统会读取文件中的数据并关闭文件。

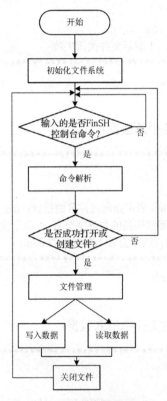

图 4.7　软件设计流程

4.2.2.3　功能设计与核心代码设计

要实现文件管理，需要使用 RT-Thread 提供的文件管理接口。本节使用 FinSH 控制台命令实现文件管理，需要自定义 FinSH 控制台命令，并将已实现的命令添加到 FinSH 控制台命令列表中，这样就可以在 MobaXterm 串口终端 FinSH 控制台中使用自定义的命令了。

1）主函数（zonesion\app\main.c）

主函数未实现任何功能，仅作为主线程。代码如下：

```
/**********************************************************************************
* 名称：main()
```

```
*********************************************************************/
int main(void)
{
    return 0;
}
```

2）w25qxx 初始化函数（zonesion\common\DRV\dev_w25qxx\dev_w25qxx.c）

见 4.1.2.3 节。

3）注册块设备函数和自动挂载列表（zonesion\common\DRV\dev_w25qxx）

见 4.1.2.3 节。

4）自定义 FinSH 控制台命令函数（zonesion\app\finsh_cmd\finsh_ex.c）

自定义 FinSH 控制台命令函数的主要工作是实现文件管理命令，包括文件的打开或创建、向文件中写入数据、从文件中读取数据、关闭文件、删除文件等，可在 FinSH 控制台中使用这些命令来实现文件管理。代码如下：

```
/*********************************************************************
 * 名称：dfsHandle (int argc, char **argv)
 * 功能：自定义 FinSH 控制台命令
 * 参数：argc 表示参数个数；argv 表示指向参数字符串指针的数组指针
 *********************************************************************/
#include <rtthread.h>
#include "dev_w25qxx.h"
//新版 RT-Thread 的 open()函数等需要引用下面两个头文件，用于替换 dfs_posix.h
#include <unistd.h>
#include <dfs.h>

int dfsHandle(int argc, char **argv)
{
    static int file = -1;
    rt_err_t result = RT_EOK;
    if(!rt_strcmp(argv[1], "open"))
    {
        file = open("/text.txt", O_RDWR | O_CREAT);        //打开或创建 text.txt 文件
        if(file >= 0)
            rt_kprintf("Open or create success! file name: text.txt\n");
        else
            rt_kprintf("Open or create fail!\n");
    }
    else if(!rt_strcmp(argv[1], "close"))
    {
        result = close(file);                              //关闭文件
        if(result >= 0)
            rt_kprintf("Close success! file name: text.txt\n");
        else
            rt_kprintf("Close fail!\n");
    }
    else if(!rt_strcmp(argv[1], "read"))
```

```
    {
        char readBuf[64] = {0};
        result = read(file, readBuf, sizeof(readBuf));         //从文件中读取数据（需要先打开文件）
        if(result > 0)
            rt_kprintf("Read %d bytes of data,data: %s\n", result, readBuf);
        else if(result == 0)
            rt_kprintf("No data!\n");
        else
            rt_kprintf("Read fail!\n");
        result = close(file);                                  //关闭文件
        if(result >= 0)
            rt_kprintf("Close success! file name: text.txt\n");
        else
            rt_kprintf("Close fail!\n");
    }
    else if(!rt_strcmp(argv[1], "write"))
    {
        char writeBuf[64] = "Hello Rt-Thread!";
        result = write(file, (char*)writeBuf, sizeof(writeBuf)); //向文件中写入数据（需要先打开文件）
        if(result >= 0)
            rt_kprintf("Write %d bytes of data,data: %s\n", result, writeBuf);
        else
            rt_kprintf("Write fail!\n");
        result = close(file);                                  //关闭文件
        if(result >= 0)
            rt_kprintf("Close success! file name: text.txt\n");
        else
            rt_kprintf("Close fail!\n");
    }
    else if(!rt_strcmp(argv[1], "unlink"))
    {
        result = unlink("/text.txt");                          //删除文件（需要先创建并关闭文件）
        if(result == 0)
            rt_kprintf("Unlink \"/text.txt\" success!\n");
        else
            rt_kprintf("Unlink fail!\n");
    }
    else if(!rt_strcmp(argv[1], "help"))
    {
        rt_kprintf("open open file\n");
        rt_kprintf("close close file\n");
        rt_kprintf("read read file\n");
        rt_kprintf("write write file\n");
        rt_kprintf("unlink unlink file\n");
    }
    else
        rt_kprintf("Please input 'dfsHandle <help|open|close|read|write|unlink>'\n");
    return 0;
```

```
}
MSH_CMD_EXPORT(dfsHandle, dfs file management);
```

4.2.3　开发步骤与验证

4.2.3.1　硬件部署

同 1.2.3.1 节。

4.2.3.2　工程调试

同 2.1.3.2 节。

4.2.3.3　验证效果

（1）关闭 RT-Thread Studio，拔掉仿真器，按下 ZI-ARMEmbed 上的电源按键重新上电。

（2）ZI-ARMEmbed 重启后，在 MobaXterm 串口终端 FinSH 控制台输出 "read W25Qxx ID is: 0x16EF"（文件系统被挂载到的设备）、"Find a Winbond flash chip. Size is 8388608 bytes."（找到 Winbond 闪存芯片并显示大小）、"W25Q64 flash device is initialize success."（W25QXX 芯片初始化成功）、"Probe SPI flash W25Q64 by SPI device spi10 success."（SPI 设备 spi10 成功探测到 SPI 闪存 W25QXX）。

```
     \ | /
- RT -       Thread Operating System
 / | \       4.1.0 build Nov   1 2022 17:27:56
 2006 - 2022 Copyright by RT-Thread team
read W25Qxx ID is: 0x16EF
[I/SFUD] Find a Winbond flash chip. Size is 8388608 bytes.
[I/SFUD] W25QXX flash device is initialize success.
[I/SFUD] Probe SPI flash W25QXX by SPI device spi10 success.
```

（3）在 MobaXterm 串口终端 FinSH 控制台中输入 "dfsHandle helps"，解析自定义 FinSH 控制台命令后，FinSH 控制台将输出可以实现的操作。

```
     \ | /
- RT -       Thread Operating System
 / | \       4.1.0 build Nov   1 2022 17:27:56
 2006 - 2022 Copyright by RT-Thread team
read W25Qxx ID is: 0x16EF
[I/SFUD] Find a Winbond flash chip. Size is 8388608 bytes.
[I/SFUD] W25QXX flash device is initialize success.
[I/SFUD] Probe SPI flash W25QXX by SPI device spi10 success.
msh /> dfsHandle helps
Please input 'dfsHandle <help|open|close|read|write|unlink>'
msh />
```

（4）在 MobaXterm 串口终端 FinSH 控制台中输入 "dfsHandle open"，解析自定义 FinSH 控制台命令后可打开或创建一个文件，FinSH 控制台将输出文件打开或创建成功的信息，并显示文件名称。

```
      \ | /
    - RT -     Thread Operating System
     / | \       4.1.0 build Nov   1 2022 17:27:56
     2006 - 2022 Copyright by RT-Thread team
    read W25Qxx ID is: 0x16EF
    [I/SFUD] Find a Winbond flash chip. Size is 8388608 bytes.
    [I/SFUD] W25QXX flash device is initialize success.
    [I/SFUD] Probe SPI flash W25QXX by SPI device spi10 success.
    msh />dfsHandle helps
    Please input 'dfsHandle <help|open|close|read|write|unlink>'
    msh />dfsHandle open
    Open or create success! file name: text.txt
    msh />
```

（5）在 MobaXterm 串口终端 FinSH 控制台中输入"dfsHandle write"，解析自定义 FinSH
控制台命令后可向文件中写入数据（注意数据写入完后关闭文件），FinSH 控制台将输出写
入数据的长度和数据、关闭文件成功的信息，并显示关闭文件的名称。

```
      \ | /
    - RT -     Thread Operating System
     / | \       4.1.0 build Nov   1 2022 17:27:56
     2006 - 2022 Copyright by RT-Thread team
    read W25Qxx ID is: 0x16EF
    [I/SFUD] Find a Winbond flash chip. Size is 8388608 bytes.
    [I/SFUD] W25QXX flash device is initialize success.
    [I/SFUD] Probe SPI flash W25QXX by SPI device spi10 success.
    msh />dfsHandle helps
    Please input 'dfsHandle <help|open|close|read|write|unlink>'
    msh />dfsHandle open
    Open or create success! file name: text.txt
    msh />dfsHandle write
    Write 64 bytes of data,data: Hello Rt-Thread!
    Close success! file name: text.txt
    msh />
```

（6）在 MobaXterm 串口终端 FinSH 控制台中输入"dfsHandle read"，解析自定义 FinSH
控制台命令后可从文件中读取数据。由于在向文件中写入数据后将文件关闭了，因此从文件
中读取数据会失败，FinSH 控制台将输出读取数据失败和关闭文件失败的信息。

```
      \ | /
    - RT -     Thread Operating System
     / | \       4.1.0 build Nov   1 2022 17:27:56
     2006 - 2022 Copyright by RT-Thread team
    read W25Qxx ID is: 0x16EF
    [I/SFUD] Find a Winbond flash chip. Size is 8388608 bytes.
    [I/SFUD] W25QXX flash device is initialize success.
    [I/SFUD] Probe SPI flash W25QXX by SPI device spi10 success.
    msh />dfsHandle helps
    Please input 'dfsHandle <help|open|close|read|write|unlink>'
```

```
msh />dfsHandle open
Open or create success! file name: text.txt
msh />dfsHandle write
Write 64 bytes of data,data: Hello Rt-Thread!
Close success! file name: text.txt
msh />dfsHandle read
Read fail!
Close fail!
msh />
```

（7）在 MobaXterm 串口终端 FinSH 控制台中输入"dfsHandle open"，解析自定义 FinSH 控制台命令后可再次打开文件或创建一个新的文件，FinSH 控制台将输出文件打开或创建成功的信息，并显示文件名称。

```
\|/
- RT -     Thread Operating System
 /|\     4.1.0 build Nov  1 2022 17:27:56
 2006 - 2022 Copyright by RT-Thread team
read W25Qxx ID is: 0x16EF
[I/SFUD] Find a Winbond flash chip. Size is 8388608 bytes.
[I/SFUD] W25QXX flash device is initialize success.
[I/SFUD] Probe SPI flash W25QXX by SPI device spi10 success.
msh />dfsHandle helps
Please input 'dfsHandle <help|open|close|read|write|unlink>'
msh />dfsHandle open
Open or create success! file name: text.txt
msh />dfsHandle write
Write 64 bytes of data,data: Hello Rt-Thread!
Close success! file name: text.txt
msh />dfsHandle read
Read fail!
Close fail!
msh />dfsHandle open
Open or create success! file name: text.txt
msh />
```

（8）在 MobaXterm 串口终端 FinSH 控制台中输入"dfsHandle read"，解析自定义 FinSH 控制台命令后可从文件中读取数据，此时文件没有被关闭，处于打开状态，从文件中读取数据成功，FinSH 控制台将输出读取数据的长度和数据、关闭文件成功的信息，并显示关闭的文件名称。

```
\|/
- RT -     Thread Operating System
 /|\     4.1.0 build Nov  1 2022 17:27:56
 2006 - 2022 Copyright by RT-Thread team
read W25Qxx ID is: 0x16EF
[I/SFUD] Find a Winbond flash chip. Size is 8388608 bytes.
[I/SFUD] W25QXX flash device is initialize success.
[I/SFUD] Probe SPI flash W25QXX by SPI device spi10 success.
```

```
msh />dfsHandle helps
Please input 'dfsHandle <help|open|close|read|write|unlink>'
msh />dfsHandle open
Open or create success! file name: text.txt
msh />dfsHandle write
Write 64 bytes of data,data: Hello Rt-Thread!
Close success! file name: text.txt
msh />dfsHandle read
Read fail!
Close fail!
msh />dfsHandle open
Open or create success! file name: text.txt
msh />dfsHandle read
Read 64 bytes of data,data: Hello Rt-Thread!
Close success! file name: text.txt
```

4.2.4　小结

本节主要介绍文件系统的常用操作和管理方式。通过本节的学习，读者可掌握 RT-Thread 文件管理接口的应用。

4.3 目录管理应用开发

在操作系统中，目录管理是指对文件系统中目录结构进行的组织、创建、删除、遍历等操作。操作系统使用一种层次化的结构来组织目录，形成目录树。每个节点可以是文件或者其他目录，而根节点是整个文件系统的根目录。

本节的要求如下：

- ➲ 了解目录管理的常用操作。
- ➲ 掌握目录管理的方式。
- ➲ 掌握 RT-Thread 目录管理接口的应用。

4.3.1　原理分析

4.3.1.1　目录管理

目录管理是操作系统的一个关键组成部分，它涉及组织和管理文件系统中的目录结构，以便应用程序可以轻松地查找、创建、访问和操作文件。目录管理的常用操作如下：

（1）目录结构：目录管理涉及文件系统的组织结构，包括目录和子目录的层次结构。不同的操作系统使用不同的目录结构，如树状结构、扁平结构、多层结构等。

（2）目录操作：应用程序可以执行各种目录操作，包括目录的创建、删除、重命名、移动和复制等。

（3）文件和目录操作：目录管理允许应用程序执行文件和目录操作，包括文件的创建、删除、复制、移动、重命名和访问等。

（4）路径系统：操作系统使用路径来标识文件和目录的位置。路径可以是绝对路径（从根目录开始的完整路径）或相对路径（相对于当前工作目录的路径）。

（5）目录权限：目录管理也包括设置目录的访问权限，以控制哪些用户或应用程序可以访问和修改目录中的文件。

（6）特殊目录：操作系统通常包括一些特殊目录，如（root）根目录、家（home）目录、临时目录等。

（7）搜索和索引：许多操作系统提供搜索和索引功能，以便用户可以快速查找文件和目录。

4.3.1.2　RT-Thread 的目录管理方式

RT-Thread 提供了一系列基本的目录操作，包括目录的创建、打开、读取、关闭等，这些操作可有效地组织和管理文件。本节主要介绍与目录操作相关的函数，对目录的操作一般都是基于目录地址进行的。RT-Thread 的常用目录操作如图 4.7 所示。

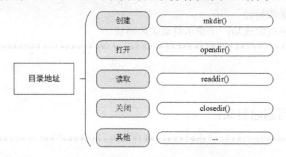

图 4.8　RT-Thread 的常用目录操作

4.3.1.3　目录管理方式

1）创建目录

通过下面的函数可创建目录：

```
/**********************************************************************
 * 名称：mkdir()
 * 功能：创建目录
 * 参数：path 表示目录的绝对地址；mode 表示创建模式
 * 返回：0 表示创建目录成功；-1 表示创建目录失败
 **********************************************************************/
int mkdir(const char *path, mode_t mode);
```

2）打开目录

通过下面的函数可打开目录：

```
/**********************************************************************
 * 名称：opendir()
 * 功能：打开目录
 * 参数：name 表示目录的绝对地址
 * 返回：指向目录的指针表示打开目录成功；NULL 表示打开目录失败
 **********************************************************************/
DIR* opendir(const char* name);
```

3）读取目录

通过下面的函数可读取目录：

```
/**********************************************************************************
 * 名称：readdir()
 * 功能：读取目录
 * 参数：d 表示目录流指针
 * 返回：指向目录的结构体指针表示读取目录成功；NULL 表示已读到目录结尾
 **********************************************************************************/
struct dirent* readdir(DIR *d);
```

4）关闭目录

通过下面的函数可关闭目录：

```
/**********************************************************************************
 * 名称：closedir()
 * 功能：关闭目录
 * 参数：d 表示目录流指针
 * 返回：0 表示目录关闭成功；-1 表示目录关闭错误
 **********************************************************************************/
int closedir(DIR* d);
```

5）删除目录

通过下面的函数可删除目录：

```
/**********************************************************************************
 * 名称：rmdir()
 * 功能：删除目录
 * 参数：pathname 表示待删除目录的绝对路径
 * 返回：0 表示目录删除成功；-1 表示目录删除错误
 **********************************************************************************/
int rmdir(const char *pathname);
```

4.3.2 开发设计与实践

4.3.2.1 硬件设计

本节的硬件设计同 3.2.2.1 节。本节通过 MobaXterm 串口终端 FinSH 控制台来控制虚拟文件系统，有助于读者掌握目录管理的应用。

4.3.2.2 软件设计

软件设计流程如图 4.9 所示。

（1）初始化文件系统。

（2）在 MobaXterm 串口终端 FinSH 控制台中输入"dfsHandle helps"，FinSH 控制台将输出"Please input 'dfsHandle <help|mkdir|rmdir|opendir|closedir>'"。

（3）在 MobaXterm 串口终端 FinSH 控制台中输入"dfsHandle mkdir"，可创建一个目录。

（4）在 MobaXterm 串口终端 FinSH 控制台中输入"dfsHandle opendir"，可打开创建的目录。

（5）在 MobaXterm 串口终端 FinSH 控制台中输入"dfsHandle closedir"，可关闭打开的目录。

（6）在 MobaXterm 串口终端 FinSH 控制台中输入"dfsHandle rmdir"，可删除指定的目录。

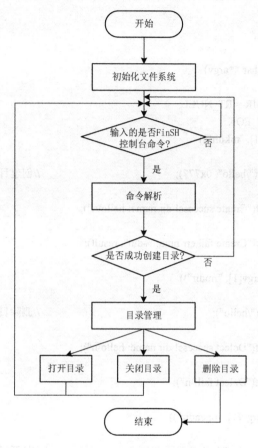

图 4.9　软件设计流程

4.3.2.3　功能设计与核心代码设计

要实现目录管理，需要使用 RT-Thread 提供的目录管理接口。本节使用 FinSH 控制台命令实现目录管理，需要自定义 FinSH 控制台命令，并将已实现的命令添加到 FinSH 控制台命令列表中，这样就可以在 MobaXterm 串口终端 FinSH 控制台中使用自定义的命令了。

1）主函数（zonesion\app\main.c）

同 4.2.2.3 节。

2）w25qxx 初始化函数（zonesion\common\DRV\dev_w25qxx\dev_w25qxx.c）

见 4.1.2.3 节。

3）注册块设备函数和自动挂载列表（zonesion\common\DRV\dev_w25qxx）

见 4.1.2.3 节。

4）自定义 FinSH 控制台命令函数（zonesion\app\finsh_cmd\finsh_ex.c）

自定义 FinSH 控制台命令函数的主要工作是实现目录管理命令，包括目录的创建、打开、删除和关闭等，可在 FinSH 控制台中使用这些命令来实现目录管理。代码如下：

```
/*****************************************************************************
 * 名称：dfsHandle (int argc, char **argv)
 * 功能：自定义 FinSH 控制台命令
 * 参数：argc 表示参数个数；argv 表示指向参数字符串指针的数组指针
 *****************************************************************************/
#include <rtthread.h>
#include "dev_w25qxx.h"
#include <dfs_posix.h>
int dfsHandle(int argc, char **argv)
{
    static DIR * tempDIR = RT_NULL;
    rt_err_t result = RT_EOK;
    if(!rt_strcmp(argv[1], "mkdir"))
    {
        result = mkdir("/hello", 0x777);                        //创建目录
        if(result == 0)
            rt_kprintf("Create success! dir name: hello\n");
        else
            rt_kprintf("Create fail,err num: %d\n", result);
    }
    else if(!rt_strcmp(argv[1], "rmdir"))
    {
        result = rmdir("/hello");                               //删除目录
        if(result == 0)
            rt_kprintf("Delect success! dir name: hello\n");
        else
            rt_kprintf("Delect fail!\n");
    }
    else if(!rt_strcmp(argv[1], "opendir"))
    {
        tempDIR = opendir("/hello");                            //打开目录
        if(tempDIR != RT_NULL)
            rt_kprintf("Open dir success!\n");
        else
            rt_kprintf("Open dir fail!\n");
    }
    else if(!rt_strcmp(argv[1], "closedir"))
    {
        result = closedir(tempDIR);                            //关闭目录
        if(result == 0)
            rt_kprintf("Close dir success!\n");
        else
            rt_kprintf("Close dir fail!\n");
    }
    else if(!rt_strcmp(argv[1], "help"))
    {
        rt_kprintf("mkdir create directory\n");
        rt_kprintf("rmdir delete directory\n");
```

```
            rt_kprintf("opendir open directory\n");
            rt_kprintf("closedir close directory\n");
        }
        else
            rt_kprintf("Please input 'dfsHandle <help|mkdir|rmdir|opendir|closedir>'\n");
        return 0;
    }
    MSH_CMD_EXPORT(dfsHandle, dfs directory management);
```

4.3.3　开发步骤与验证

4.3.3.1　硬件部署

同 1.2.3.1 节。

4.3.3.2　工程调试

同 2.1.3.2 节。

4.3.3.3　验证效果

（1）关闭 RT-Thread Studio，拔掉仿真器，按下 ZI-ARMEmbed 上的电源按键重新上电。

（2）ZI-ARMEmbed 重启后，在 MobaXterm 串口终端 FinSH 控制台输出 "read W25Qxx ID is: 0x16EF"（文件系统被挂载到的设备）、"Find a Winbond flash chip. Size is 8388608 bytes."（找到 Winbond 闪存芯片并显示大小）、"W25Q64 flash device is initialize success."（W25QXX 芯片初始化成功）、"Probe SPI flash W25Q64 by SPI device spi10 success."（SPI 设备 spi10 成功探测到 SPI 闪存 W25QXX）。

```
 \ | /
- RT -     Thread Operating System
 / | \     4.1.0 build Nov   2 2022 11:00:28
 2006 - 2022 Copyright by RT-Thread team
read W25Qxx ID is: 0x16EF
[I/SFUD] Find a Winbond flash chip. Size is 8388608 bytes.
[I/SFUD] W25QXX flash device is initialize success.
[I/SFUD] Probe SPI flash W25QXX by SPI device spi10 success.
msh />
```

（3）在 MobaXterm 串口终端 FinSH 控制台中输入 "dfsHandle helps"，解析自定义 FinSH 控制台命令后，FinSH 控制台将输出可以实现的操作。

```
 \ | /
- RT -     Thread Operating System
 / | \     4.1.0 build Nov   2 2022 11:00:28
 2006 - 2022 Copyright by RT-Thread team
read W25Qxx ID is: 0x16EF
[I/SFUD] Find a Winbond flash chip. Size is 8388608 bytes.
[I/SFUD] W25QXX flash device is initialize success.
[I/SFUD] Probe SPI flash W25QXX by SPI device spi10 success.
msh />dfsHandle helps
```

```
Please input 'dfsHandle <help|mkdir|rmdir|opendir|closedir>'
msh />
```

（4）在 MobaXterm 串口终端 FinSH 控制台中输入"dfsHandle mkdir"，解析自定义 FinSH 控制台命令后可创建目录，FinSH 控制台将输出目录创建成功信息，并显示目录名字。

```
 \ | /
 - RT -     Thread Operating System
 / | \      4.1.0 build Nov   2 2022 11:00:28
 2006 - 2022 Copyright by RT-Thread team
read W25Qxx ID is: 0x16EF
[I/SFUD] Find a Winbond flash chip. Size is 8388608 bytes.
[I/SFUD] W25QXX flash device is initialize success.
[I/SFUD] Probe SPI flash W25QXX by SPI device spi10 success.
msh />dfsHandle helps
Please input 'dfsHandle <help|mkdir|rmdir|opendir|closedir>'
msh />dfsHandle mkdir
Create success! dir name: hello
msh />
```

（5）在 MobaXterm 串口终端 FinSH 控制台中输入"dfsHandle opendir"，解析自定义 FinSH 控制台命令后可打开目录，FinSH 控制台将输出目录打开成功的信息。

```
 \ | /
 - RT -     Thread Operating System
 / | \      4.1.0 build Nov   2 2022 11:00:28
 2006 - 2022 Copyright by RT-Thread team
read W25Qxx ID is: 0x16EF
[I/SFUD] Find a Winbond flash chip. Size is 8388608 bytes.
[I/SFUD] W25QXX flash device is initialize success.
[I/SFUD] Probe SPI flash W25QXX by SPI device spi10 success.
msh />dfsHandle helps
Please input 'dfsHandle <help|mkdir|rmdir|opendir|closedir>'
msh />dfsHandle mkdir
Create success! dir name: hello
msh />dfsHandle opendir
Open dir success!
msh />
```

（6）在 MobaXterm 串口终端 FinSH 控制台中输入"dfsHandle closedir"，解析自定义 FinSH 控制台命令后可关闭目录，FinSH 控制台将输出目录关闭成功的信息。

```
 \ | /
 - RT -     Thread Operating System
 / | \      4.1.0 build Nov   2 2022 11:00:28
 2006 - 2022 Copyright by RT-Thread team
read W25Qxx ID is: 0x16EF
[I/SFUD] Find a Winbond flash chip. Size is 8388608 bytes.
[I/SFUD] W25QXX flash device is initialize success.
[I/SFUD] Probe SPI flash W25QXX by SPI device spi10 success.
```

```
msh />dfsHandle helps
Please input 'dfsHandle <help|mkdir|rmdir|opendir|closedir>'
msh />dfsHandle mkdir
Create success! dir name: hello
msh />dfsHandle opendir
Open dir success!
msh />dfsHandle closedir
Close dir success!
msh />
```

（7）在 MobaXterm 串口终端 FinSH 控制台中输入 "dfsHandle rmdir"，解析自定义 FinSH 控制台命令后可删除目录，FinSH 控制台将输出目录删除成功的信息，并显示删除目录的名称。

```
 \ | /
- RT -     Thread Operating System
 / | \     4.1.0 build Nov  2 2022 11:00:28
 2006 - 2022 Copyright by RT-Thread team
read W25Qxx ID is: 0x16EF
[I/SFUD] Find a Winbond flash chip. Size is 8388608 bytes.
[I/SFUD] W25QXX flash device is initialize success.
[I/SFUD] Probe SPI flash W25QXX by SPI device spi10 success.
msh />dfsHandle helps
Please input 'dfsHandle <help|mkdir|rmdir|opendir|closedir>'
msh />dfsHandle mkdir
Create success! dir name: hello
msh />dfsHandle opendir
Open dir success!
msh />dfsHandle closedir
Close dir success!
msh />dfsHandle rmdir
Delect success! dir name: hello
msh />
```

4.3.4　小结

本节主要介绍目录的常用操作和管理方式。通过本节的学习，读者可掌握 RT-Thread 目录管理接口的应用。

第 5 章
RT-Thread GUI 开发技术

5.1 GUI 基础和 emWin 图形库应用开发

嵌入式 GUI（Graphical User Interface）是专为嵌入式系统设计的图形用户界面，可以使嵌入式设备能够展示用户友好的图形界面，支持交互和信息显示。嵌入式 GUI 在很多领域都得到了广泛的应用，如医疗设备、工业控制、消费电子、汽车信息娱乐系统等。

本节的要求如下：
- 了解 LCD 的工作原理。
- 了解 emWin 图形库的特性及使用。
- 掌握 emWin 的管理方式。
- 掌握 emWin 管理接口的应用。

5.1.1 原理分析

5.1.1.1 LCD 的工作原理

在引入图形库概念之前，本节先简要介绍 LCD 的工作原理。液晶显示器（Liquid Crystal Display，LCD）广泛应用在电子设备的平面显示中，其工作原理涉及液晶材料的光学特性、电场的作用，以及通过控制液晶分子的排列方向来实现图像显示的过程。

LCD 的结构是在两片平行的玻璃基板中间放置液晶盒，下玻璃基板上设置了 TFT（薄膜晶体管），上玻璃基板上设置了彩色滤光片，通过改变 TFT 上的信号与电压可以控制液晶分子的排列方向，从而控制每个像素的偏振光，最终达到显示的目的。液晶显示的工作原理是液晶分子在不同电压的作用下会呈现不同的光特性。

在使用各种 GUI 之前，需要先完成 LCD 的驱动（正常显示文字、图形等），再完成 LCD 层、GUI 层的设置与适配，从而在屏幕上显示基于图形库的各种控件等。GUI 及驱动如图 5.1 所示。

5.1.1.2 emWin 简介

emWin 是 SEGGER 公司针对嵌入式系统开发的稳定、高效的图形软件库，适用于图形操作应用，并可输出高质量、无锯齿的文字和图形，通过调用 emWin 提供的函数，可以使嵌入式图形界面的开发变得简单而快捷。emWin 特点如下：

（1）图形功能：emWin 支持各种图形功能，包括基本的图形（如线条、矩形、圆形等）绘制、图形显示、文本渲染等，提供了丰富的 API 来进行图形操作。

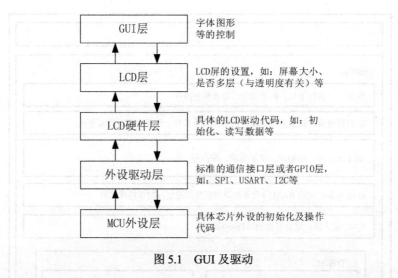

图 5.1　GUI 及驱动

（2）用户界面（UI）：emWin 不仅提供了图形功能，还可以构建复杂的用户界面。通过 emWin 创建按钮、对话框、滑块、列表框等 UI 元素，开发者可以构建交互性强的应用程序。

（3）支持多种显示设备：emWin 能够适应不同的显示设备，包括 LCD、电子墨水屏、OLED 等，可支持不同的显示分辨率和颜色深度。

（4）支持跨平台：emWin 不依赖于特定的操作系统，可用于不同的嵌入式系统，如裸机系统、RTOS（实时操作系统）等。

（5）支持触摸屏：emWin 支持触摸屏，可用于触摸屏的交互式应用，提供了触摸屏驱动程序和相关的 API，简化了触摸屏的开发过程。

emWin 在医疗设备、工控设备、汽车信息娱乐系统等领域得到了广泛的应用，为开发用户界面提供了丰富的工具和库。这些工具和库可以实现以下功能：

➥ 绘制 2D 图形：绘制圆、椭圆、多边形、弧线、线图和饼形图等。

➥ 显示图像：显示 BMP、JPEG、GIF 及 PNG 等格式的图像。

➥ 显示文字：显示多种文字。

➥ 处理用户输入：可处理诸如键盘、鼠标及触摸屏等的输入。

➥ 提供各种窗口对象：这些窗口对象也称为图形控件，如菜单控件、窗口控件、按钮控件、复选框控件和框架窗口控件等，基于这些控件可以非常容易地制作类似 Windows 风格的界面。

5.1.1.3　STemWin 图形库

STemWin 和 emWin 实际上是同一个图形库，只是在命名上有所不同。STemWin 是 ST 公司对 emWin 图形库的一种定制和移植，专门用于 STM32 系列微控制器。ST 公司将 emWin 作为 STM32 系列微控制器的官方图形库，并将 emWin 移植并集成到 STM32Cube 软件工具套件中。

5.1.1.4　emWin 的架构

emWin 的架构如图 5.2 所示。

图 5.2　emWin 的架构图

5.1.1.5　emWin 图形库管理方式

1）初始化 emWin

在使用 emWin 前，需要先初始化 emWin 内部的数据结构和变量。RT-Thread 提供的 GUI_Init()函数会调用 LCD_X_Config()函数来打开 LCD 设备。通过下面的函数可初始化 emWin：

```
/*****************************************************************************
 * 名称：GUI_Init()
 * 功能：初始化 emWin 内部数据结构和变量
 * 返回：0 表示成功；其他值表示初始化失败
 *****************************************************************************/
int GUI_Init(void);
```

2）窗口启用内存设备

通过下面的函数可令窗口启用内存设备：

```
/*****************************************************************************
 * 名称：WM_SetCreateFlags()
 * 功能：在创建新窗口时设置标志
 * 参数：Flags 表示窗口创建标志
 * 返回：此参数以前的值
 *****************************************************************************/
U32　WM_SetCreateFlags(U32　Flags);
```

3）启用 UTF-8 编码

通过下面的函数可启用 UTF-8 编码：

```
/*******************************************************************
 * 名称：GUI_UC_SetEncodeUTF8()
 * 功能：启用 UTF-8 编码，emWin 会根据 UTF-8 转换规则对给定字符串进行编码
 *******************************************************************/
void    GUI_UC_SetEncodeUTF8(void);
```

4）执行所有挂起的 emWin 任务

在 emWin 中，通过 GUI_Exec() 函数可执行 emWin 任务。该函数通常是在没有操作系统的情况下使用的。在带有操作系统的情况下，可使用 GUI_Delay() 函数来执行 emWin 任务。执行 GUI_Delay() 函数相当于在执行 GUI_Exec() 函数后执行一个延时函数。对于实时系统来说，一个线程不会一直占用 CPU 的资源，通过 rt_thread_mdelay() 函数可以使线程释放 CPU 的资源。GUI_Delay() 函数会调用 rt_thread_mdelay() 函数，从而挂起 emWin 任务。

```
/*******************************************************************
 * 名称：GUI_Exec() 和 GUI_Delay()
 * 功能：执行 emWin 任务
 * 返回：0 表示未执行任务；1 表示执行了一个任务
 *******************************************************************/
int GUI_Exec(void);
void GUI_Delay(int Period);
```

5）设置背景色

通过下面的函数可设置背景色：

```
/*******************************************************************
 * 名称：GUI_SetBkColor()
 * 功能：设置当前的背景色
 * 参数：Color 表示背景色，该参数是 24 位的 RGB 值
 * 返回：选定的背景色
 *******************************************************************/
void    GUI_SetBkColor(GUI_COLOR Color);
```

6）清除窗口

通过下面的函数可清除窗口：

```
/*******************************************************************
 * 名称：GUI_Clear()
 * 功能：清除指定的窗口，如果没有指定窗口，则窗口就是整个显示屏，此时整个显示屏会被清除
 *******************************************************************/
void    GUI_Clear(void);
```

7）绘制用颜色填充水平梯度的矩形

通过下面的函数可绘制用颜色填充水平梯度的矩形：

```
/*******************************************************************
 * 名称：GUI_DrawGradientH()
 * 功能：绘制用颜色填充水平梯度的矩形
 * 参数：x0 表示左上角的横轴坐标；y0 表示左上角的纵轴坐标；x1 表示右下角的横轴坐标；y1 表
示右下角的纵轴坐标；Color0 表示矩形最左侧的颜色；Color1 表示矩形最右侧的颜色
```

```
********************************************************************************/
void   GUI_DrawGradientH (int x0,int y0,int x1,int y1,GUI_COLOR   Color0,
                          GUI_COLOR   Color1);
```

5.1.2　开发设计与实践

5.1.2.1　硬件设计

本节通过编程来控制 ZI-ARMEmbed 的 LCD，将 emWin 与 LCD 的显示结合起来实现联动控制，帮助读者掌握 emWin 的使用方法。ZI-ARMEmbed 和 LCD 的连接如图 5.3 所示。

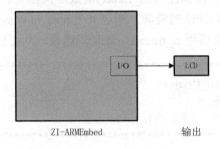

图 5.3　ZI-ARMEmbed 和 LCD 的连接

5.1.2.2　软件设计

软件设计流程如图 5.4 所示。

（1）初始化并启动 GUI 线程。

（2）初始化 emWin 并打开 LCD，在 GUI 线程 gui_thread_entry 的入口函数中完成 emWin 的初始化工作。在 emWin 的初始化中调用 LCD_X_Config()函数，查找系统中注册的 LCD，获取相应的 LCD 操作函数。

（3）通过 GUI_SetBkColor()和 GUI_Clear()函数刷新背景色，通过 GUI_DrawGradientH() 函数绘制用颜色填充水平梯度的矩形。

（4）执行 GUI_Delay()函数，刷新图形并延时。

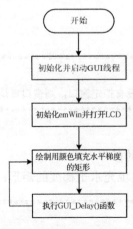

图 5.4　软件设计流程

5.1.2.3　功能设计与核心代码设计

1）主函数（zonesion\app\main.c）

主函数的主要工作是完成 GUI 线程的初始化，初始化完成之后退出主函数。代码如下：

```
/********************************************************************************
* 名称：main()
* 功能：GUI 线程初始化
********************************************************************************/
#include "gui_thread.h"
int main(void)
{
    gui_thread_init();
    return 0;
}
```

2）GUI 线程初始化函数（zonesion\app\gui_thread.c）

GUI 线程初始化函数的主要工作是创建并启动 GUI，包括定义线程名称、入口函数，设置线程内存大小、优先级、时间片等参数。代码如下：

```
/********************************************************************************
* 名称：gui_thread_init()
* 功能：创建并启动 GUI 线程
* 返回：其他表示线程启动成功；-1 表示线程启动失败
********************************************************************************/
#include "gui_thread.h"
#include "emwin/emwin_thread.h"
int gui_thread_init(void)
{
    rt_thread_t gui_thread = RT_NULL;
    gui_thread = rt_thread_create("gui",                //线程名称
                        gui_thread_entry,               //线程入口函数
                        RT_NULL,                        //线程入口函数参数
                        1024,                           //线程内存大小
                        5,                              //线程优先级
                        20);                            //线程时间片
    if(gui_thread == RT_NULL)
        return -1;
    return rt_thread_startup(gui_thread);               //启动线程
}
```

3）GUI 线程入口函数（zonesion\app\gui_thread.c）

GUI 线程入口函数的主要工作是初始化 emWin，设置背景色，在死循环中绘制用颜色填充水平梯度的矩形，并完成 GUI 图形的刷新和延时。代码如下：

```
/********************************************************************************
* 名称：gui_thread_entry (void *parameter)
* 功能：GUI 线程的入口函数
* 参数：parameter 表示入口函数参数，在线程创建时传入
********************************************************************************/
void gui_thread_entry(void *parameter)
```

```
{
    (void)parameter;
    short x = 0, y = 0;
    unsigned char num = 0;
    emwin_init();                                   //初始化 emWin
    GUI_SetBkColor(GUI_WHITE);                      //设置背景色
    GUI_Clear();                                    //清除当前窗口，刷新为背景色
    while(1)
    {
        if(x >= 320)                                //制造位移变色效果
        {
            x = -60;
            y += 20;
            if(y >= 240)
            {
                num++;
                if(num >= 3)
                    num = 0;
                y = 0;
            }
        }
        //绘制用颜色填充水平梯度的矩形
        GUI_DrawGradientH(x, y, x+60, y+20, color[num][0], color[num][1]);
        x++;
        GUI_Delay(5);                               //GUI 刷新和延时
    }
}
```

4）emWin 初始化函数（zonesion\app\gui_thread.c）

emWin 初始化函数的主要工作是初始化 emWin、设置 VM 窗口管理器的动态内存大小、开启 CRC 时钟（emWin 必须要求打开）。代码如下：

```
/*************************************************************************
* 名称：emwin_init (void)
* 功能：初始化 emWin
*************************************************************************/
void emwin_init(void)
{
    __HAL_RCC_CRC_CLK_ENABLE();                     //开启 CRC 时钟
    GUI_Init();                                     //初始化 emWin
    WM_SetCreateFlags(WM_CF_MEMDEV);                //设置 VM 窗口管理器的动态内存大小
    GUI_UC_SetEncodeUTF8();                         //使用 UTF-8 编码
}
```

5.1.3 开发步骤与验证

5.1.3.1 硬件部署

同 1.2.3.1 节。

5.1.3.2　工程调试

同 2.1.3.2 节。

5.1.3.3　验证效果

（1）关闭 RT-Thread Studio，拔掉仿真器，按下 ZI-ARMEmbed 上的电源按键重新上电。

（2）观察 ZI-ARMEmbed 上的 LCD，最开始 LCD 显示白色的背景，然后用颜色填充水平梯度，使 LCD 显示位移变色的效果，如图 5.5 所示。

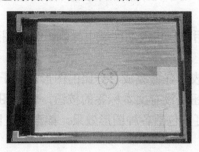

图 5.5　位移变色的效果

5.1.4　小结

本节主要介绍 LCD 的工作原理、emWin 图形库的特性及管理方式。通过本节的学习，读者可掌握 emWin 图形库接口的应用。

5.2 GUI 图形和颜色应用开发

GUI 通常不仅包括各种图形元素，如线条、矩形、圆形等，用于绘制复杂的界面和图形；还包括各种颜色。颜色在 GUI 中是一个重要的视觉元素，用于增强用户体验、区分元素、传递信息和美化界面。

本节的要求如下：

⊃ 了解 emWin 绘制图形的基本函数。

⊃ 熟悉 emWin 中 GUI 图形和颜色绘制的管理方式。

⊃ 掌握 emWin 中 GUI 图形和颜色绘制的应用。

5.2.1　原理分析

5.2.1.1　GUI 图形与颜色

1）图形

GUI 中的常用图形有：

（1）矩形和圆形：矩形和圆形是最基本的图形之一，用于构建窗口、面板和其他界面组件的外观。

（2）线条：线条用于连接点、划分区域或表示界面中的分隔线。

（3）图标：图标是小型图形，通常表示应用程序、文件或特定功能。图标可用于快速识别和启动相应的操作。

（4）按钮：按钮是包含文本或图标的交互式图形，用于触发特定的操作。按钮可以采用不同的样式和外观。

（5）图像：GUI 中的图像可以是位图、矢量图或其他格式的图形文件。图像用于显示复杂的图形内容，如图标、背景图或应用程序中的图像。

（6）文本框：文本框用于显示文本信息，并允许用户输入文本内容。文本框可以分为单行文本框和多行文本框。

（7）滚动条：滚动条是用于在可滚动区域中导航内容的图形，它允许用户在大量数据中选择感兴趣的部分。

（8）进度条：进度条是表示任务完成百分比的图形，用于显示操作的进度。

（9）图形按钮：图形按钮是没有文本标签的按钮，使用图形或图标来表示按钮的功能。

（10）图形效果：GUI 可以包括各种图形效果，如渐变、阴影、透明度等，以提高界面的视觉吸引力。

（11）图形绘图区域：可以在其中绘制自定义图形的区域，如绘图应用程序中的画布。

上述的图形共同构成了 GUI 的视觉部分，使用户能够通过直观的图形界面与计算机程序或系统进行交互。在嵌入式系统中，图形库（如 STemWin）可以用于轻松地创建和管理这些图形，提供丰富的图形功能。

2）颜色

GUI 中颜色的一些常见应用和概念：

（1）背景色：界面的整体背景色通常设置为某种中性颜色，以确保其他元素能够清晰可见。

（2）前景色：前景色通常是指文字、图标等元素的颜色，通常需要与背景色形成对比，以确保良好的可读性。

（3）按钮颜色：按钮通常具有与主题相符的颜色，以突出其交互性。激活状态、悬停状态和禁用状态的按钮可能会显示不同的颜色。

（4）链接颜色：链接通常用特殊颜色表示，以便用户能够识别并单击它们。单击后的链接颜色可能会发生变化，以表示用户已经单击过该链接。

（5）警告颜色：用于强调警告、错误或重要信息的颜色，通常是红色或其他醒目的颜色。

（6）成功颜色：用于强调成功或正面信息的颜色，通常是绿色或其他积极的颜色。

（7）进度条颜色：进度条的颜色通常用于表示任务的进展情况，不同的颜色可能表示不同的状态。

（8）图形颜色：界面中的各种图形，如线条、矩形、图标等，都可能有各自的颜色。

（9）主题颜色：一些 GUI 允许用户定义主题颜色，以便在应用程序中保持一致的外观。

（10）背景渐变和纹理：背景色可以与渐变或纹理相结合，以提供更加丰富的外观和感觉。

（11）透明度：颜色的透明度属性允许界面元素具有半透明效果，以创建更为复杂的图层和重叠效果。

5.2.1.2　emWin 的图形绘制接口

emWin 提供了大量的图形绘制接口，通过这些接口的组合调用，可实现复杂的 GUI。

基本图形绘制接口如表 5.1 所示，线条绘制接口如表 5.2 所示，多边形绘制接口如表 5.3 所示，圆、椭圆、弧线、曲线绘制接口如表 5.4 所示。

表 5.1 基本图形绘制接口

接 口 名 称	描 述	接 口 名 称	描 述
GUI_Clear()	用背景色填充屏幕/激活窗口	GUI_ClearRect()	用背景色填充一个矩形区域
GUI_CopyRect()	在屏幕上复制一个矩形区域	GUI_DrawGradientH()	绘制用颜色填充水平梯度的矩形
GUI_DrawGradientV()	绘制用颜色填充垂直梯度的矩形	GUI_DrawGradientRoundedH()	绘制用颜色填充水平梯度的圆角矩形
GUI_DrawGradientRoundedV()	绘制用颜色填充垂直梯度的圆角矩形	GUI_DrawPixel()	绘制单个像素
GUI_DrawPoint()	绘制点	GUI_DrawRect()	绘制矩形
GUI_DrawRectEx()	绘制矩形	GUI_DrawRoundedFrame()	绘制圆角框
GUI_DrawRoundedRect()	绘制圆角矩形	GUI_FillRect()	绘制用颜色填充的矩形
GUI_FillRectEx()	绘制用颜色填充的矩形	GUI_FillRoundedRect()	绘制用颜色填充的圆角矩形
GUI_InvertRect()	倒转矩形区域	—	—

表 5.2 线条绘制接口

接 口 名 称	描 述	接 口 名 称	描 述
GUI_DrawHLine()	绘制水平线	GUI_DrawLineRel()	绘制从当前位置到按距离指定的终点的线条（相对坐标）
GUI_DrawLine()	绘制从指定起点到指定终点的线条（绝对坐标）	GUI_DrawLineTo()	绘制从当前位置到指定终点的线条
GUI_DrawPolyLine()	绘制折线	GUI_GetLineStyle()	返回当前的线条样式
GUI_DrawVLine()	绘制垂直线	GUI_MoveRel()	相对于当前位置移动线条指针
GUI_MoveTo()	将线条指针移动到给定位置	GUI_SetLineStyle()	设置当前的线条样式

表 5.3 多边形绘制接口

接 口 名 称	描 述	接 口 名 称	描 述
GUI_DrawPolygon()	绘制多边形的轮廓	GUI_MagnifyPolygon()	放大多边形
GUI_EnlargePolygon()	扩展多边形	GUI_RotatePolygon()	按指定角度旋转多边形
GUI_FillPolygon()	绘制填充的多边形	—	—

表 5.4 圆、椭圆、弧线、曲线绘制接口

接 口 名 称	描 述	接 口 名 称	描 述
GUI_DrawCircle()	绘制圆的轮廓	GUI_FillEllipse()	绘制用颜色填充的椭圆
GUI_FillCircle()	绘制用颜色填充的圆	GUI_DrawArc()	绘制弧线
GUI_DrawEllipse()	绘制椭圆的轮廓	GUI_DrawGraph()	绘制曲线

5.2.1.3　emWin 的颜色绘制接口

颜色绘制的基本接口如表 5.5 所示，颜色绘制的转换接口如表 5.6 所示。

表 5.5　颜色绘制的基本接口

接 口 名 称	描　述	接 口 名 称	描　述
GUI_GetBkColor()	返回当前背景色	GUI_SetBkColor()	设置当前背景色
GUI_GetBkColorIndex()	返回当前背景色的索引	GUI_SetBkColorIndex()	设置当前背景色的索引
GUI_GetColor()	返回当前前景色	GUI_SetColor()	设置当前前景色
GUI_GetColorIndex()	返回当前前景色的索引	GUI_SetColorIndex()	设置当前前景色的索引
GUI_GetDefaultColor()	返回默认前景色	GUI_SetDefaultColor()	设置默认前景色
GUI_GetDefaultBkColor()	返回默认背景色	GUI_SetDefaultBkColor()	设置默认背景色

表 5.6　颜色绘制的转换接口

接口名称	描　述	接口名称	描　述
GUI_CalcColorDist()	返回两种颜色之间的差值	GUI_Color2VisColor()	返回最接近的可用颜色
GUI_CalcVisColorError()	将差值返回给下一种可用颜色	GUI_ColorIsAvailable()	检查给定的颜色是否可用
GUI_Color2Index()	将颜色转换为颜色索引	GUI_Index2Color()	将颜色索引转换为颜色

5.2.1.4　emWin 的图形和颜色管理方式

1）设置前景色

通过下面的函数可设置当前前景色：

```
/***************************************************************************
 * 名称：GUI_SetColor()
 * 功能：设置当前前景色
 * 参数：Color 表示当前前景色，该参数是 24 位的 RGB 值
 * 返回：选定的前景色
 ***************************************************************************/
void GUI_SetColor(GUI_COLOR Color);
```

2）绘制直线

通过下面的函数可在当前窗口中绘制直线（部分直线会由于没有位于当前窗口或者由于当前部分窗口不可见而不可见，这是由于裁减造成的）：

```
/***************************************************************************
 * 名称：GUI_DrawLine()
 * 功能：在当前窗口中绘制从某个指定起点到某个指定终点之间的线（绝对坐标）
 * 参数：x0 表示起点的横轴坐标；y0 表示起点的纵轴坐标；x1 表示终点的横轴坐标；y1 表示终点
的纵轴坐标
 ***************************************************************************/
void   GUI_DrawLine(int x0,int y0,int x1,int y1);
```

3）绘制矩形

通过下面的函数可在当前窗口中绘制矩形：

```
/*********************************************************************
 * 名称：GUI_DrawRect()
 * 功能：在当前窗口中的指定位置绘制矩形
 * 参数：x0 表示矩形左上角的横轴坐标；y0 表示矩形左上角的纵轴坐标；x1 表示矩形右下角的横轴
坐标；y1 表示矩形右下角的纵轴坐标
 *********************************************************************/
void   GUI_DrawRect(int x0,int y0,int x1,int y1);
```

4）绘制圆形

通过下面的函数可在当前窗口中绘制圆形：

```
/*********************************************************************
 * 名称：GUI_FillCircle()
 * 功能：在当前窗口中的指定位置绘制指定尺寸的圆形
 * 参数：x0 表示圆心的横轴坐标（像素）；y0 表示圆心的纵轴坐标（像素）；r 表示圆的半径（直径
的一半），必须为正数
 *********************************************************************/
void   GUI_FillCircle(int x0,int y0,int r);
```

5）绘制椭圆形

通过下面的函数可在当前窗口中绘制椭圆形：

```
/*********************************************************************
 * 名称：GUI_FillEllipse()
 * 功能：在当前窗口中的指定位置绘制指定尺寸的填充的椭圆
 * 参数：x0 表示椭圆中心的横轴坐标（像素）；y0 表示椭圆中心的纵轴坐标（像素）；rx 表示椭圆的
长半轴，必须为正数；ry 表示椭圆的短半轴，必须为正数
 *********************************************************************/
void   GUI_FillEllipse(int x0,int y0,int rx,int ry);
```

6）绘制饼形

通过下面的函数可在当前窗口中绘制饼形：

```
/*********************************************************************
 * 名称：GUI_DrawPie()
 * 功能：绘制圆形扇区
 * 参数：x0 表示饼形中心的横轴坐标（像素）；y0 表示饼形中心的纵轴坐标（像素）；r 表示圆的半
径，必须为正数；a0 表示起始角度（度）；a1 表示结束角度（度）；Type 为保留参数，应为 0
 *********************************************************************/
void   GUI_DrawPie(int x0,int y0,int r,int a0,int a1,int Type);
```

7）绘制渐变色填充的圆角矩形

通过下面的函数可在当前窗口中绘制用渐变色填充的圆角矩形：

```
/*********************************************************************
 * 名称：GUI_DrawGradientRoundedH()
 * 功能：绘制用水平颜色梯度填充的圆角矩形
 * 参数：x0 表示矩形左上角的横轴坐标；y0 表示矩形左上角的纵轴坐标；x1 表示矩形右下角的横轴
坐标；y1 表示矩形右下角的纵轴坐标；rd 表示圆角的半径；Color0 表示矩形最左侧要填充的颜色；Color1
表示矩形最右侧要填充的颜色
```

```
***********************************************************************/
void GUI_DrawGradientRoundedH(int x0,int y0, int x1,int y1,int rd,GUI_COLOR Color0,
                              GUI_COLOR Color1);
```

5.2.2　开发设计与实践

5.2.2.1　硬件设计

本节的硬件设计同 5.1.2.1 节。

5.2.2.2　软件设计

软件设计流程如图 5.6 所示。

（1）初始化并启动 GUI 线程。

（2）初始化 emWin 并打开 LCD，在 GUI 线程 gui_thread_entry 的入口函数中完成 emWin 的初始化工作。在 emWin 的初始化中调用 LCD_X_Config()函数，查找系统中注册的 LCD，获取相应的 LCD 操作函数。

（3）绘制直线、矩形、圆形、椭圆形、饼形、圆角矩形。

（4）进入 while 循环，执行 GUI_Delay()函数。

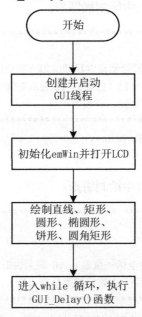

图 5.6　软件设计流程

5.2.2.3　功能设计与核心代码设计

1）主函数（zonesion\app\main.c）

同 5.1.2.3 节。

2）GUI 线程初始化函数（zonesion\app\gui_thread.c）

同 5.1.2.3 节。

3）GUI 线程入口函数（zonesion\app\gui_thread.c）

GUI 线程入口函数的主要工作是初始化 emWin，设置背景色，绘制直线、矩形、圆形、椭圆形、饼形、圆角矩形。代码如下：

```
/************************************************************************
* 名称：gui_thread_entry (void *parameter)
* 功能：GUI 线程入口函数
* 参数：parameter 表示入口函数参数，在创建 GUI 线程时传入
************************************************************************/
void gui_thread_entry(void *parameter)
{
    (void)parameter;
    int a0 = 0, a1 = 0;
    emwin_init();                                        //初始化 emWin
    GUI_SetBkColor(GUI_WHITE);                           //设置背景色
    GUI_Clear();                                         //清除当前窗口，刷新背景色
    GUI_SetColor(GUI_RED);                               //设置颜色为红色
    GUI_DrawLine(20, 20, 80, 70);                        //在指定位置绘制直线
    GUI_SetColor(GUI_GREEN);                             //设置颜色为绿色
    GUI_DrawRect(100, 20, 150, 70);                      //在指定位置绘制矩形
    GUI_SetColor(GUI_BLUE);                              //设置颜色为蓝色
    GUI_FillCircle(200, 40, 30);                         //在指定位置绘制圆形
    GUI_SetColor(GUI_LIGHTMAGENTA);                      //设置颜色为洋红色
    GUI_FillEllipse(270, 40, 25, 30);                    //在指定位置绘制椭圆形
    for(int i=0; i<GUI_COUNTOF(value); i++)              //绘制饼形
    {
        a0 = (i == 0) ? 0 : value[i - 1];
        a1 = value[i];
        GUI_SetColor(color[i]);
        GUI_DrawPie(100, 150, 50, a0, a1, 0);            //绘制扇形
    }
    //在指定位置绘制用渐变色填充的圆角矩形
    GUI_DrawGradientRoundedH(190, 120, 260, 180, 5, 0x0000FF, 0x00FFFF);
    while(1){
        GUI_Delay(10);                                   //GUI 刷新和延时
    }
}
```

4）emWin 初始化函数（zonesion\app\gui_thread.c）

同 5.1.2.3 节。

5.2.3 开发步骤与验证

5.2.3.1 硬件部署

同 1.2.3.1 节。

5.2.3.2 工程调试

同 2.1.3.2 节。

5.2.3.3　验证效果

（1）关闭 RT-Thread Studio，拔掉仿真器，按下 ZI-ARMEmbed 上的电源按键重新上电。

（2）ZI-ARMEmbed 上的 LCD 显示绘制的红色直线、绿色矩形、蓝色圆形、洋红色椭圆形、扇形、用渐变色填充的圆角矩形。绘制的图形如图 5.7 所示。

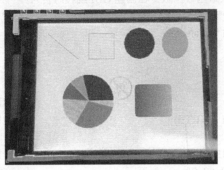

图 5.7　绘制的图形

5.2.4　小结

本节主要介绍 emWin 的图形与颜色的绘制原理和管理方式。通过本节的学习，读者可掌握 emWin 图形和颜色绘制接口的应用。

5.3　GUI 文本显示应用开发

GUI 文本显示是指在图形用户界面中显示文本内容。文本在 GUI 中广泛用于标签、按钮、菜单、文本框、对话框、状态栏等元素，用于向用户提供信息、提示、标签和控制选项。GUI 文本显示是创建用户友好、信息丰富且可定制的图形用户界面的基础。

本节的要求如下：

- ⊃ 熟悉 emWin 文本显示的管理方式。
- ⊃ 掌握 emWin 文本显示接口的应用。

5.3.1　原理分析

5.3.1.1　GUI 文本显示

GUI 文本显示可以包括各种元素，如文本标签、按钮、菜单项、文本框、列表框等，用于传递信息、提供选项和用户输入。以下是关于 GUI 文本显示的重要概念：

（1）文本标签（Text Labels）：文本标签用于在 GUI 中显示静态文本，如标题、说明、标签等。它们通常用于表示 GUI 元素的用途。

（2）按钮（Buttons）：按钮通常包括文本，用于显示与按钮相关联的操作。按钮文本可以是标签或动词性的文本，如"保存""取消"等。

（3）文本框（Text Fields）：文本框用于输入文本，它们通常包含占位文本，用于指导用户输入所需的信息。

（4）列表框（List Boxes）：列表框用于显示一系列项目，通常是文本，用户可以从中选择具体的项目。列表框可以是单选列表框或多选列表框。

（5）文本样式和格式：文本在 GUI 中通常有不同的样式和格式，如字体、颜色、大小、粗细等，这样可以增强文本的可读性和吸引力。

（6）多语言支持：许多应用程序和操作系统都支持多语言，因此 GUI 文本显示需要支持多种语言和字符集，包括使用 Unicode 等字符编码方案。

（7）动态文本：有时 GUI 需要显示动态文本或实时更新文本，如系统状态、实时日志、消息通知等。

（8）布局和对齐：GUI 文本显示的布局和对齐对 GUI 的外观和用户体验而言是至关重要的。文本可以左对齐、右对齐、居中对齐等，也可以在不同的布局容器中排列。

（9）国际化和本地化：对于支持全球用户群体的应用程序，GUI 文本显示需要考虑国际化和本地化，以适应不同文化和地区的需求。

（10）文本输入验证：在文本输入框中，可以添加验证和格式检查，以确保用户输入的数据是有效的且符合要求。

（11）屏幕阅读器兼容性：对于对可访问性要求较高的应用程序，需要考虑屏幕阅读器的兼容性，确保文本可以被正确解读和表达。

5.3.1.2　emWin 文本显示

emWin 不仅支持字符串、数字等文本的显示，还支持文本位置、进制、编码、字体等设置。emWin 提供了功能强大且灵活的文本显示功能，使得其在嵌入式系统中呈现文本变得相对简单。以下是使用 emWin 文本显示的一般步骤：

（1）字体选择：emWin 支持多种字体和字号，在开始使用 emWin 时，需要选择适合应用的字体。字体通常是以文件的形式存储的，如 TrueType 字体文件。

（2）字体的加载与配置：如果需要使用外部字体，则需要将相应的字体加载到 emWin 中。emWin 提供了相应的 API 来加载和配置字体。

（3）文本显示：emWin 提供了用于文本显示的函数，包括单行文本和多行文本的显示。使用 GUI_DispStringAt()函数可以在指定位置显示文本。

（4）文本颜色和背景色：emWin 允许设置文本颜色和背景色。通过 GUI_SetColor()函数可以设置文本颜色，通过 GUI_SetBkColor()函数可以设置背景色。

（5）文本对齐和格式：emWin 提供了文本对齐和格式的控制接口。通过 GUI_SetTextMode()和 GUI_SetTextAlign()等函数可以设置文本的对齐方式和格式。

（6）文本显示区域设置：emWin 允许设置文本显示的区域，以适应特定的应用场景。通过 GUI_SetClipRect()函数可以设置文本的显示区域。

上述步骤是基本步骤。通过上述步骤，开发人员可在嵌入式系统中使用 emWin 进行文本显示。根据具体的应用场景和需求，开发人员还可以使用 emWin 提供的与文本显示相关的其他 API 来实现更复杂的文本操作。emWin 的文档提供了详细的说明和示例，对于开发者来说是一个很好的参考资料。

5.3.1.3　文本显示接口

文本显示接口如表 5.7 所示。

表 5.7　文本显示接口

接口名称	描述	接口名称	描述
GUI_DispCEOL()	清除从当前位置到行末的文本	GUI_DispStringHCenterAt()	在指定位置水平居中显示字符串
GUI_DispChar()	在当前位置显示单个字符	GUI_DispStringInRect()	在指定的矩形区域中显示字符串
GUI_DispCharAt()	在指定位置显示单个字符	GUI_DispStringInRectEx()	在指定的矩形区域中显示旋转的字符串
GUI_DispChars()	按指定数量显示字符	GUI_DispStringInRectWrap()	在指定的矩形区域中显示自动换行的字符串
GUI_DispString()	在当前位置显示字符串	GUI_DispStringInRectWrapEx()	在指定的矩形区域中显示旋转的和自动换行的字符串
GUI_DispStringAt()	在指定位置显示字符串	GUI_DispStringLen()	在当前位置显示指定数量的字符串
GUI_DispStringAtCEOL()	在指定位置显示字符串，并清除从指定位置到行末的文本	GUI_WrapGetNumLines()	返回使用自动换行模式在指定区域显示指定字符串所需的行号

5.3.1.4　数值显示接口

数值显示接口如表 5.8 所示。

表 5.8　数值显示接口

接口名称	描述	接口名称	描述
GUI_DispDec()	在当前位置显示指定字符数的十进制数值	GUI_DispFloatFix()	显示指定小数点右边位数的浮点数值
GUI_DispDecAt()	在指定位置显示指定字符数的十进制数值	GUI_DispFloatMin()	显示最小字符数的浮点数值
GUI_DispDecMin()	在当前位置显示最小字符数的十进制数值	GUI_DispSFloatFix()	显示指定小数点右边位数的浮点数值并显示符号
GUI_DispDecShift()	在当前位置显示指定字符数、带小数点的十进制数值	GUI_DispSFloatMin()	显示最小字符数的浮点数值并显示符号
GUI_DispDecSpace()	在当前位置显示指定字符数的十进制数值，用空格代替首位的 0	GUI_DispBin()	在当前位置显示二进制数值
GUI_DispSDec()	在当前位置显示指定字符数的十进制数值并显示符号	GUI_DispBinAt()	在指定位置显示二进制数值
GUI_DispSDecShift()	在当前位置显示指定字符数、带小数点的十进制数值并显示符号	GUI_DispHex()	在当前位置显示十六进制数值
GUI_DispFloat()	显示指定字符数的浮点数值	GUI_DispHexAt()	在指定位置显示十六进制数值

5.3.1.5　字体设置接口

字体设置接口如表 5.9 所示。

表 5.9　字体设置接口

接口名称	描述	接口名称	描述
GUI_GetCharDistX()	返回当前字体中指定字符的宽度像素	GUI_GetTextExtend()	评估使用当前字体的文本大小

GUI_GetDefaultFont()	返回默认字体	GUI_GetTrailingBlankCols()	返回给定字符的后导空格像素列数
GUI_GetFont()	返回当前选择的字体的指针	GUI_GetYDistOfFont()	返回特定字体的高度间距
GUI_GetFontDistY()	返回当前字体的高度间距	GUI_GetYSizeOfFont()	返回特定字体的高度
GUI_GetFontInfo()	返回包含字体信息的结构	GUI_IsInFont()	评估特定字体中是否存在指定的字符
GUI_GetFontSizeY()	返回当前字体的高度像素	GUI_SetDefaultFont()	设置 GUI_Init()后使用的默认字体
GUI_GetLeadingBlankCols()	返回给定字符的前导空格像素列数	GUI_SetFont()	设置当前字体
GUI_GetStringDistX()	返回使用当前字体的文本的宽度	—	—

5.3.1.6 emWin 文本显示管理方式

1）设置文本输入位置

通过下面的函数可设置文本输入位置：

```
/*********************************************************************
 * 名称：GUI_GotoXY()
 * 功能：设置文本写入位置
 * 参数：x 表示横轴坐标（单位为像素，0 表示左边界）；y 表示纵轴坐标（单位为像素，0 表示顶部
边界）
 * 返回：0 表示成功；不等于 0 表示当前文本位置超出窗口范围，这样后续的写入操作可能被忽略
 *********************************************************************/
char   GUI_GotoXY (int x,int y);
```

2）设置字体

通过下面的函数可设置字体：

```
/*********************************************************************
 * 名称：GUI_SetFont()
 * 功能：设置用于文本输入的字体
 * 参数：pFont 表示指向所选字体的指针
 * 返回：指向所选字体的指针，以便进行缓存
 *********************************************************************/
Const   GUI_FONT *   GUI_SetFont(const GUI_FONT * pNewFont);
```

3）显示字符

通过下面的函数可在当前窗口中显示单个字符（是否显示字符取决于所选择的字体，如果在当前字体中该字符不可用，则不会显示该字符）：

```
/*********************************************************************
 * 名称：GUI_DispChar()
 * 功能：在当前窗口的当前文本位置处，使用当前字体显示单个字符
 * 参数：c 表示要显示的字符
 *********************************************************************/
void   GUI_DispChar (U16 c);
```

4）显示十进制数值

通过下面的函数可显示十进制数值：

```
/*******************************************************************************
 * 名称：GUI_DispDec()
 * 功能：在当前窗口的当前文本位置处，使用当前字体显示指定字符数的十进制数值
 * 参数：v 表示显示的数值，最小值为-2147483648，最大值为 2147483647；Len 表示显示的位数（最
大为 10）
 *******************************************************************************/
void    GUI_DispDec (I32 v,U8 Len);
```

5）显示字符串

通过下面的函数可在当前窗口中显示字符串（字符串可以包括控制字符串 "\n"）：

```
/*******************************************************************************
 * 名称：GUI_DispString()
 * 功能：在当前窗口的当前文本位置处，使用当前字体显示作为参数的字符串
 * 参数：s 表示要显示的字符串
 *******************************************************************************/
void    GUI_DispString (const char *s);
```

5.3.2　开发设计与实践

5.3.2.1　硬件设计

本节的硬件设计同 5.1.2.1 节。

5.3.2.2　软件设计

软件设计流程如图 5.8 所示。

（1）初始化并启动 GUI 线程。

（2）初始化 emWin 并打开 LCD，在 GUI 线程 gui_thread_entry 的入口函数中完成 emWin 的初始化工作。在 emWin 的初始化中调用 LCD_X_Config()函数，查找系统中注册的 LCD，获取相应的 LCD 操作函数。

（3）设置颜色、位置、字体。

（4）显示字符、数值、字符串。

（5）进入 while 循环，执行 GUI_Delay()函数。

图 5.8　软件设计流程

5.3.2.3　功能设计与核心代码设计

1）主函数（zonesion\app\main.c）

同 5.1.2.3 节。

2）GUI 线程初始化函数（zonesion\app\gui_thread.c）

同 5.1.2.3 节。

3）GUI 线程入口函数（zonesion\app\gui_thread.c）

GUI 线程入口函数的主要工作是初始化 emWin，设置背景色、前景色、显示位置、字体，并显示字符、数值、字符串。系统在 GUI 线程启动之后调用 GUI 线程入口函数，GUI 线程入口函数执行完毕后不再进入该入口函数。代码如下：

```
/*****************************************************************************
* 名称：gui_thread_entry (void *parameter)
* 功能：GUI 线程入口函数
* 参数：parameter 表示入口函数参数，在创建 GUI 线程时传入
*****************************************************************************/
void gui_thread_entry(void *parameter)
{
    (void)parameter;
    emwin_init();
     //初始化 emWin
    GUI_SetBkColor(GUI_WHITE);                              //设置背景色
    GUI_Clear();                                           //清除当前窗口，刷新背景色
    GUI_SetColor(GUI_BLACK);                               //设置当前前景色
    GUI_GotoXY(10, 10);                                    //跳转到指定位置
    GUI_SetFont(&GUI_Font8x8);                             //设置当前字体
    for(unsigned char c='A'; c<='Z'; c++)
    {
        GUI_DispChar(c);                                   //在当前位置显示字符
        GUI_Delay(500);                                    //GUI 刷新和延时
    }
    GUI_SetColor(GUI_RED);                                 //设置当前前景色
    GUI_GotoXY(10, 30);                                    //跳转到指定位置
    GUI_SetFont(&GUI_Font8x16);                            //设置当前字体
    for(unsigned char i=0; i<10; i++)
    {
        GUI_DispDec(i, 1);                                 //在当前位置显示数值
        GUI_Delay(500);                                    //GUI 刷新和延时
    }
    GUI_SetFont(&GUI_Fontsong16);                          //设置当前字体
    GUI_SetColor(GUI_DARKMAGENTA);                         //设置当前前景色
    GUI_GotoXY(10, 60);                                    //跳转到指定位置
    GUI_DispString("欢迎使用 RT-Thread！ ");               //显示字符串
    while(1){
        GUI_Delay(10);                                     //GUI 刷新和延时
    }
}
```

4）emWin 初始化函数（zonesion\app\gui_thread.c）

同 5.1.2.3 节。

5.3.3　开发步骤与验证

5.3.3.1　硬件部署

同 1.2.3.1 节。

5.3.3.2　工程调试

同 2.1.3.2 节。

5.3.3.3　验证效果

（1）关闭 RT-Thread Studio，拔掉仿真器，按下 ZI-ARMEmbed 上的电源按键重新上电。

（2）ZI-ARMEmbed 的 LCD 在指定位置显示字符，第一行以 500 ms 的时间间隔显示字母 A～Z；在指定位置显示数值，第二行以 500 ms 的时间间隔显示 1～9；在指定位置显示字符串，第三行显示"欢迎使用 RT-Thread!"。字符、数值和字符串的显示效果如图 5.9 所示。

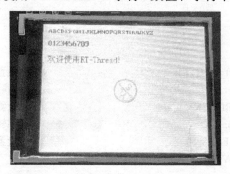

图 5.9　字符、数值和字符串的显示效果

5.3.4　小结

本节主要介绍 emWin 的文本显示的原理和管理方式。通过本节的学习，读者可掌握 emWin 文本显示接口的应用。

5.4 GUI 图像显示应用开发

在 GUI 中，图像显示是一种重要的元素，可用于丰富用户界面、传递信息和提高视觉吸引力。GUI 图像显示可显示各种图像，如图标、照片、图形、按钮图像等，这对于创建吸引人的、信息丰富的 GUI 非常重要。emWin 当前支持 BMP、JPEG、GIF 等格式的图像。

本节的要求如下：

➲ 学习 emWin 图像显示的管理方式。

➲ 掌握 emWin 图像显示接口的应用。

5.4.1 原理分析

5.4.1.1 GUI 图像显示

GUI 图像显示的关键概念如下:

(1)图像控件:GUI 图像显示通常是通过图像控件或图像框架来实现的,这些图像控件或图像框架允许开发者将图像插入到 GUI 界面中的特定位置。

(2)图标和按钮图像:图标和按钮图像用于图形按钮、菜单项、工具栏按钮等,通常表示操作、功能或导航。

(3)图像格式:图像可以采用不同的格式,如 JPEG、PNG、GIF、BMP 等。图像格式取决于图像的特性和应用程序的要求。

(4)图像大小和分辨率:图像的大小和分辨率可根据 GUI 的设计和显示设备进行调整。通常,不同的分辨率和屏幕尺寸需要不同大小的图像。

(5)透明度:图像可以具有透明背景,这使得图像可以融入 GUI 并覆盖其他元素,从而实现特定的视觉效果。

(6)动画:GUI 可以显示动画或动态图像,如旋转加载指示器、动画图标等。

(7)图像效果:图像可以设置各种效果,如阴影、光晕、颜色变换等,以提高视觉吸引力。

(8)图像单击操作:在一些 GUI 中,用户可以单击图像执行操作,如打开文件、导航到链接或执行其他功能。

5.4.1.2 emWin 图像显示接口

emWin 支持图像资源的加载,将图像直接显示在 GUI 上,表 5.10 列出了位图绘制接口。

表 5.10 位图绘制接口

接口名称	描述
GUI_DrawBitmap()	绘制位图
GUI_DrawBitmapEx()	绘制可缩放的位图
GUI_DrawBitmapHWAlpha()	在具有硬件 Alpha 混合的 GUI 上绘制带 Alpha 混合信息的位图
GUI_DrawBitmapMag()	绘制放大的位图

5.4.1.3 emWin 图像显示的管理方式

通过下面的函数可以在当前窗口的指定位置绘制位图:

```
/*******************************************************************************
 * 名称:GUI_DrawBitmap()
 * 功能:在当前窗口中的指定位置绘制位图
 * 参数:pBM 表示指向要显示的位图的指针;x 表示位图左上角在当前窗口中的横轴坐标;y 表示位
图左上角在当前窗口中的纵轴坐标
 ******************************************************************************/
void   GUI_DrawBitmap (const GUI_BITMAP * pBM, int x, int y);
```

5.4.2 开发设计与实践

5.4.2.1 硬件设计

同 5.1.2.1 节。

5.4.2.2 软件设计

软件设计流程如图 5.10 所示。

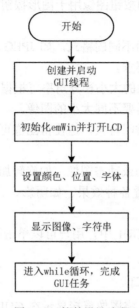

图 5.10 软件设计流程

5.4.2.3 功能设计与核心代码设计

要实现 GUI 图像显示，首先要对 LCD 进行初始化，然后使用 emWin 提供的 API 初始化 GUI。

1）主函数（zonesion\app\main.c）

同 5.1.2.3 节。

2）GUI 线程初始化函数（zonesion\app\gui_thread.c）

同 5.1.2.3 节。

3）GUI 线程入口函数（zonesion\app\gui_thread.c）

GUI 线程入口函数的主要工作是初始化 emWin，设置颜色、位置、字体，在设置完成后显示图像、字符串。代码如下：

```
/****************************************************************************
* 名称：gui_thread_entry (void *parameter)
* 功能：GUI 线程入口函数
* 参数：parameter 表示入口函数参数，在线程创建时传入
****************************************************************************/
void gui_thread_entry(void *parameter)
{
```

```
(void)parameter;
emwin_init();                              //设置背景色
GUI_Clear();                               //清除当前窗口，刷新背景色
GUI_DrawBitmap(&bmfavicon, 128, 40);       //在指定位置显示图像
GUI_SetFont(&GUI_Font8x16);                //设置字体
GUI_GotoXY(128, 115);                      //跳转到指定位置
GUI_SetColor(GUI_BLACK);                   //设置前景色
GUI_DispString("Zonesion");                //在当前位置显示字符串
GUI_DrawBitmap(&bmrtt, 57, 135);           //在指定位置显示图像
while(1){
    GUI_Delay(10);                         //GUI 刷新并延时
}
}
```

4）emWin 初始化函数（zonesion\app\gui_thread.c）

同 5.1.2.3 节。

5.4.3 开发步骤与验证

5.4.3.1 硬件部署

同 1.2.3.1 节。

5.4.3.2 项目部署

同 2.1.3.2 节。

5.4.3.3 验证效果

（1）关闭 RT-Thread Studio，拔掉仿真器，按下 ZI-ARMEmbed 上的电源按键重新上电。

（2）ZI-ARMEmbed 的 LCD 会在指定位置绘制位图，如 RT-Thread 的图标，并在指定位置显示字符串，如 Zonesion（见图 5.11）。

图 5.11 在指定位置绘制位图并显示字符串

5.4.4 小结

本节主要介绍 emWin 的图像显示的管理方式。通过本节的学习，读者可掌握 emWin 图像显示接口的应用。

5.5 GUI 控件应用开发

控件是具有对象类型属性的窗口，是组成图形用户界面的元素。控件可自动对某些事件做出反应；例如，按下某按钮后，它可以不同状态显示。和窗口一样，控件可通过其创建函数返回的句柄进行引用。

控件要求使用窗口管理器。控件创建后，可像其他任何窗口一样进行操作，窗口管理器确保了在必要时随时正确显示（并重绘）它。

本节的要求如下：

- ➲ 熟悉窗口管理器的基本原理和回调机制与方法。
- ➲ 学习 emWin 中 GUI 控件的管理方式。
- ➲ 掌握 emWin 中 GUI 控件管理接口的应用。

5.5.1　原理分析

5.5.1.1　GUI 控件

GUI 控件是用于构建和呈现图形用户界面的可视化元素，允许用户与应用程序进行交互，执行各种任务。常见的 GUI 控件如下：

（1）文本标签（Label）：用于在 GUI 中显示静态文本，如标题、说明和标签。文本标签通常用于标识其他控件或提供文本信息。

（2）按钮（Button）：允许用户触发与按钮相关联的操作，如"确定"按钮、"取消"按钮等。

（3）文本框（Text Box）：用于允许用户输入文本数据。文本框包括单行文本框和多行文本框。

（4）列表框（List Box）：用于显示一列选项，用户可以从中选择某一选项。列表框可以是单选列表框或多选列表框。

（5）下拉列表框（Combo Box）：结合了文本框和列表框，允许用户选择项目或输入自定义文本。

（6）单选按钮（Radio Button）：用户可以从一组互斥的选项中选择一个。

（7）复选框（Check Box）：复选框是独立的，用户可以选择或取消多个选项。

（8）滑块（Slider）：允许用户在指定范围内选择一个值，通常用于设置参数或进行参数调整。

（9）进度条（Progress Bar）：用于显示任务的完成进度，通常用于显示文件下载、安装等的进度。

（10）菜单（Menu）：提供应用程序选项和功能列表，通常分为菜单栏和弹出菜单。

（11）工具栏（Tool Bar）：包含快速访问功能和操作按钮。

5.5.1.2 窗口管理器

使用 emWin 的窗口管理器（WM）时，屏幕上出现的任何内容都包含在窗口中（屏幕上的一个矩形区域）。窗口尺寸可以自定义，屏幕上可以一次显示多个窗口，甚至部分或整个窗口可以在其他窗口的前面。窗口管理器提供一组函数，利用这些函数不仅可以很容易地对窗口进行创建、移动、调整大小等操作，还能操控任意数量的窗口。

5.5.1.3 窗口管理器的回调机制与方法

窗口管理器可以与回调例程一起使用，也可以不使用回调例程。在大多数情况下，最好使用回调例程。emWin 为窗口和窗口对象提供回调例程的机制（回调机制）背后的思想是事件驱动系统。如同在大多数窗口中一样，控制流不仅是从用户程序到图形系统，而且是通过用户程序提供的回调例程从用户程序到图形系统再到用户程序。

通过 GUI_Exec() 或 GUI_Delay() 函数可以完成窗口函数的回调。在窗口管理器处理消息时，大部分的 GUI 代码是在回调函数中执行的，应用程序可以通过消息接口将消息发送到窗口管理器，从而完成窗口的切换。

1）使用回调渲染

要想在窗口中使用回调例程，则需要一个回调函数，在创建窗口时将该回调函数作为 WM_CreateWindows() 函数的一部分。回调函数的原型如下：

```
/**********************************************************************************
 * 名称：void Callback(WM_MESSAGE * pMsg);
 * 功能：窗口的回调函数
 * 参数：pMsg 表示指向类型 WM_MESSAGE 的数据结构指针
 * 返回：无
 **********************************************************************************/
void Callback(WM_MESSAGE * pMsg);
```

回调函数执行的操作取决于其收到的消息类型。回调函数原型后通常跟一条 switch 语句，使用一条或多条 case 语句为不同的消息定义不同的操作，但至少要有一条处理 WM_PAINT 消息的 case 语句。当非透明窗口（默认情况下窗口是非透明窗口）收到 WM_PAINT 消息时，窗口管理器确保该消息已被选中，并重绘窗口的无效区域。最简单的方式是重绘窗口的整个区域。窗口管理器的裁剪机制可确保仅重绘窗口的无效区域。为了加速绘制过程，仅重绘窗口的无效区域是非常有用的。本节后续将介绍如何获得窗口的无效区域。

2）重写回调函数

在 emWin 中，控件和窗口的默认行为是在其回调函数中定义的。如果要更改控件的行为或者增强窗口的功能以满足客户的需求，则建议重写回调函数。重写回调函数的步骤如下：

（1）使用下面的函数创建自定义的回调函数：

```
void Callback(WM_MESSAGE * pMsg);
```

（2）实现对消息的响应。由于自定义的回调函数不需要处理所有可能的消息，所以建议使用 switch-case 语句来添加或删除某条消息的特定代码（不会影响其他消息）。参数 pMsg

包含了消息的 ID（pMsg->MsgId）。在 emWim 的官方手册中可查看窗口管理器能够处理的消息。

（3）使用内部的回调函数处理未定义的消息，推荐的方法是使用 switch-case 语句中的 default 语句来调用内部的回调函数。控件的内部回调函数是<WIDGET>_Callback()，其他类型窗口的内部回调函数是 WM_DefaultProc()。

（4）通过 WM_SetCallback()设置窗口使用的自定义的回调函数。

3）背景窗口重绘和回调

窗口管理器在初始化期间会创建一个包含整个 LCD 屏幕的窗口，并将其作为背景窗口，该窗口句柄为 WM_HBKWIN。窗口管理器不会自动重绘背景窗口的区域，因为没有默认的背景色。也就是说，如果创建了一个窗口然后将其删除，则删除的窗口仍然可见，需要通过 WM_SetDesktopColor()设置背景窗口的背景色。通过回调函数也可以处理背景窗口的重绘问题，在创建一个窗口后将其删除，回调函数将触发窗口管理器来确认背景窗口不再有效，并自动重绘背景窗口。

4）窗口无效化

对窗口或窗口的一部分进行无效化，相当于告诉窗口管理器在下次调用 GUI_Exec()或 GUI_Delay()时重绘窗口的无效区域。emWin 的窗口无效化函数不会重绘窗口的无效部分，只管理窗口的无效区域。

（1）窗口的无效区域。对于每个窗口，窗口管理器只使用一个矩形来存储包含整个无效区域的最小矩形。如果左上角的一部分和右下角的一部分变为无效，则整个窗口变为无效的。

（2）使用窗口无效化的原因。使用窗口无效化而不是立即绘制每个窗口的优点是只需要绘制一次，即使多次进行窗口无效化。例如，当更改窗口的若干属性（如窗口的背景色、字体和大小）时，在每个属性更改后立即绘制窗口比在所有属性更改后仅绘制一次窗口要花费更多的时间。

（3）窗口无效区域的重绘。通过向窗口发送一条或多条 WM_PAINT 消息，可调用函数 GUI_Exec()重绘窗口的无效区域。

5.5.1.4　控件（窗口对象）

控件是具有对象类型属性的窗口，是组成用户界面的元素。控件可自动对某些事件做出反应，如按下某个按钮后，该按钮可显示不同的状态。控件在存续期间可随时修改其属性，当不再使用某个控件时可删除该控件。

控件要求使用窗口管理器。在创建控件后，窗口管理器可确保在必要时随时正确显示并重绘控件。虽然在应用程序和用户界面中不强制使用控件，但使用控件可缩短开发时间。

1）可用控件

常用的控件如表 5.11 所示。

表 5.11　常用的控件

控 件 名 称	屏幕截图（经典）	屏幕截图（皮肤设置）	描　　述
BUTTON	Button	Button	可单击的按钮，按钮上可显示文本或位图
CHECKBOX	☑	☑ Checkbox	可选中或取消选中的复选框

续表

控 件 名 称	屏幕截图（经典）	屏幕截图（皮肤设置）	描　　述
DROPDOWN			下拉列表，单击后会打开一个列表框
EDIT	Edit	—	单行文本框，提示用户输入数字或文本
FRAMEWIN	Window	Window	框架窗口，可创建典型的 GUI
GRAPH		—	曲线控件，可显示曲线或测量值
HEADER	Red　Green　Blue	Item#　EAN ▼　Amount ☺	标题控件，用于管理各项目

2）理解重绘机制

在调用 WM_Exec()、GUI_Exec()或 GUI_Delay()后，控件可根据其属性绘制自身。在多任务环境中，通常由后台任务来调用 WM_Exec()并更新控件，在更新控件的属性后，该控件的窗口（或部分窗口）会被标记为无效，但不会立即重绘控件。重绘操作由窗口管理器稍后执行，也可以调用 WM_Paint()来强制执行重绘操作或调用 WM_Exec()来重绘所有的窗口。

3）如何使用控件

例如，使用下面的代码可以显示一个进度条：

```
PROGBAR_Handle hProgBar;
GUI_DispStringAt("Progress bar", 100, 20);
hProgBar = PROGBAR_Create(100, 40, 100, 20, WM_CF_SHOW);
```

上面的第一行代码用来为控件（进度条）设置句柄，最后一行代码用来创建进度条。在任务中调用 WM_Exec()时，窗口管理器会自动绘制进度条。

每种类型的控件都有若干能够修改控件外观的成员函数。在创建控件后，通过调用控件的成员函数可更改控件的属性，这些成员函数将控件句柄作为第一个参数。例如，要使上面代码创建的进度条显示 45%，并将进度条的颜色从默认的颜色（通常为深灰/浅灰）更改为绿色/红色，可使用以下代码：

```
PROGBAR_SetBarColor(hProgBar, 0, GUI_GREEN);
PROGBAR_SetBarColor(hProgBar, 1, GUI_RED);
PROGBAR_SetValue(hProgBar, 45);
```

使用控件时应注意：

（1）默认配置。控件具有一个或多个配置宏定义，这些宏定义有不同的默认设置，如所使用的字体和颜色。

（2）控件通信。控件通常是作为子窗口创建的，父窗口可以是任何类型的窗口，甚至还可以是另一种控件。为了确保同步，无论任何子窗口有任何事件发生，通常都应通知父窗口。

当事件发生时，子窗口通过发送 WM_NOTIFY_PARENT 消息与其父窗口通信。通知代码取决于事件类型，通常作为消息的一部分发送到父窗口。

5.5.1.5　emWin 控件管理方式

1）创建对话框

通过下面的函数可创建对话框：

```
/*********************************************************************
 * 名称：GUI_CreateDialogBox()
 * 功能：创建对话框
 * 参数：paWidget 表示指向对话框包含的控件资源表的指针；NumWidgets 表示对话框中包含的控件
总数；cb 表示指向回调函数（对话框过程函数）的指针；hParent 表示父窗口句柄（0 表示没有父窗口），通
常以桌面为父窗口；x0 表示对话框相对于父窗口的水平位置坐标；y0 表示对话框相对于父窗口的垂直位置
坐标
 * 返回：对话框句柄，该句柄可用于访问资源表中的第一个控件
 *********************************************************************/
WM_HWIN   GUI_CreateDialogBox(const GUI_WIDGET_CREATE_INFO * paWidget,
                    int NumWidgets,WM_CALLBACK * cb, WM_HWIN hParent,int x0, int y0);
```

2）设置 MULITEDIT 控件显示为空

通过下面的函数可设置 MULITEDIT 控件显示为空：

```
/*********************************************************************
 * 名称：MULTIEDIT_SetText()
 * 功能：设置将由 MULTIEDIT 控件处理的文本
 * 参数：hObj 表示 MULTIEDIT 控件句柄；s 表示指向 MULTIEDIT 控件待处理的文本的指针
 *********************************************************************/
void   MULTIEDIT_SetText(MULTIEDIT_HANDLE hObj, const char *s);
```

3）为 MULITEDIT 控件添加字符

通过下面的函数可以为 MULITEDIT 控件添加文本：

```
/*********************************************************************
 * 名称：MULTIEDIT_AddText()
 * 功能：在当前光标位置添加指定的文本
 * 参数：hObj 表示 MULTIEDIT 控件句柄；s 表示指向待添加文本的指针
 *********************************************************************/
int   MULTIEDIT_AddText(MULTIEDIT_HANDLE   hObj, const char *s);
```

5.5.2　开发设计与实践

5.5.2.1　硬件设计

本节通过 MobaXterm 串口终端 FinSH 控制台来控制 LCD 的显示，实现 FinSH 控制台、ZI-ARMEmbed 和 LCD 的联动。FinSH 控制台、ZI-ARMEmbed 和 LCD 的连接如图 5.12 所示。

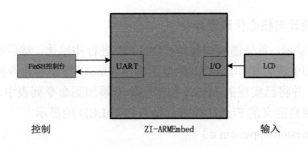

图 5.12　FinSH 控制台、ZI-ARMEmbed 和 LCD 的连接

5.5.2.2　软件设计

软件设计流程如图 5.13 所示。

（1）在主函数中创建并启动 GUI 线程。

（2）在线程入口函数中完成 emWin 的初始化并打开 LCD。

（3）在线程入口函数中创建对话框。

（4）在 MobaXterm 串口终端 FinSH 控制台中输入"guiCtrl show"命令并附带一个参数，可调用 MULTIEDIT_AddText()函数为 MULITEDIT 控件添加字符，在 LCD 上显示相应的字符。

（5）在 MobaXterm 串口终端 FinSH 控制台中输入"guiCtrl clear"命令，可调用 MULTIEDIT_SetText()函数设置 MULITEDIT 控件显示为空，清除 LCD 上显示的所有字符。

（6）在 while 循环中执行 GUI_Delay()函数，该函数可调用窗口管理器的消息处理回调函数_cbDialog()。

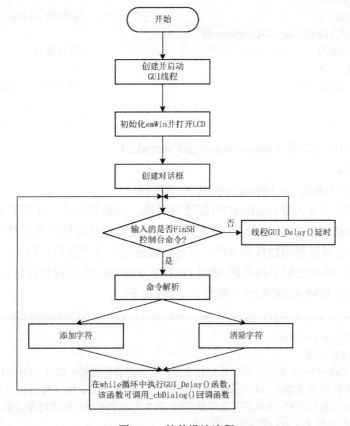

图 5.13　软件设计流程

5.5.2.3　功能设计与核心代码设计

要控制 LCD 的显示，首先需要编写程序对 LCD 进行初始化，然后初始化 emWin，并使用 emWin 提供的 API 进行操作。LCD 的显示需由 FinSH 控制台命令控制，因此需要实现 FinSH 控制台命令，并将已实现的 FinSH 控制台命令添加到命令列表中，这样在启动 FinSH 控制台后就可以使用自定义的 FinSH 控制台命令控制 LCD 的显示。

1）主函数（zonesion\app\main.c）

同 5.1.2.3 节。

2）GUI 线程初始化函数（zonesion\app\gui_thread.c）

同 5.1.2.3 节。

3）GUI 线程入口函数（zonesion\app\gui_thread.c）

GUI 线程入口函数首先实现 emWin 初始化，并创建窗口，然后不断执行 GUI_Delay() 函数来完成 GUI 任务的回调执行。代码如下：

```
/*********************************************************************
 * 名称：gui_thread_entry (void *parameter)
 * 功能：GUI 线程入口函数
 * 参数：parameter 表示入口函数参数，在创建 GUI 线程时传入
 *********************************************************************/
void gui_thread_entry(void *parameter)
{
    (void)parameter;
    emwin_init();                                      //初始化 emWin
    extern WM_HWIN CreateGUIctrl(void);
    CreateGUIctrl();                                   //创建窗口
    while(1){
        GUI_Delay(10);                                 //GUI 刷新并延时
    }
}
```

4）emWin 初始化函数（zonesion\app\gui_thread.c）

同 5.1.2.3 节。

5）创建对话框函数（zonesion\app\GUIctrlDLG.c）

使用函数 GUI_CreateDialogBox()可创建对话框，该函数包含一个回调函数 _cbDialog()，并使用了_aDialogCreate 控件列表。也就是说，在创建对话框时就包含了控件列表中的控件。使用控件的好处是可以使用控件的 API，简化后续应用程序的编写。例如，下面的代码可创建了一个对话框，其中使用了控件的 MULTIEDIT_SetText()来完成控件字体的显示；所创建对话框的父窗口为 WM_HBKWIN（桌面）。代码如下：

```
/*********************************************************************
 * 名称：GUI_CreateDialogBox()
 * 功能：创建对话框
 * 参数：_aDialogCreate 表示指向对话框包含的控件资源表的指针；GUI_COUNTOF(_aDialogCreate)
表示对话框中包含的控件总数；_cbDialog 表示指向应用程序回调函数（对话框过程函数）的指针；
WM_HBKWIN 表示父窗口句柄,这里的父窗口是桌面；第 1 个 0 表示对话框相对于父窗口的水平位置坐标；
第 2 个 0 表示对话框相对于父窗口的垂直位置坐标
```

```
*******************************************************************************/
#include "DIALOG.h"
WM_HWIN CreateGUIctrl(void) {
    WM_HWIN hWin;
    hWin = GUI_CreateDialogBox(_aDialogCreate, GUI_COUNTOF(_aDialogCreate), _cbDialog,
                               WM_HBKWIN, 0, 0);
    return hWin;
}
```

6）自定义 FinSH 控制台命令函数（zonesion/app/finsh_cmd/finsh_ex.c）

自定义 FinSH 控制台命令函数用来控制 LCD 的显示，代码如下：

```
/*******************************************************************************
* 名称：guiCtrl (int argc, char **argv)
* 功能：自定义 FinSH 控制台命令
* 参数：argc 表示参数的个数；argv 表示指向参数字符串指针的数组指针
*******************************************************************************/
#include <rtthread.h>
#include <stdio.h>
#include <stdlib.h>
void Multi_addStr(char *str);
int guiCtrl(int argc, char **argv)
{
    if(!rt_strcmp(argv[1], "show"))
    {
        if(argc < 3)                                        //检测命令是否包含参数
        {
            rt_kprintf("Please enter a display string!\n");
            return 0;
        }
        char tempBuf[128] = {0};
        snprintf(tempBuf, sizeof(tempBuf), "%s\n", argv[2]);
        Multi_addStr(tempBuf);                              //在 MULITEDIT 中添加字符
    }
    else if(!rt_strcmp(argv[1], "clear"))
    {
        Multi_addStr(RT_NULL);                              //清除 MULITEDIT 中的字符
    }
    else
        rt_kprintf("Please input 'guiCtrl <show|clear>'\n");
    return 0;
}
MSH_CMD_EXPORT(guiCtrl, GUI control sample);
```

5.5.3　开发步骤与验证

5.5.3.1　硬件部署

同 1.2.3.1 节。

5.5.3.2　工程调试

同 2.1.3.2 节。

5.5.3.3　验证效果

（1）关闭 RT-Thread Studio，拔掉仿真器，按下 ZI-ARMEmbed 上的电源按键重新上电。

（2）在 MobaXterm 串口终端 FinSH 控制台中输入"guiCtrl show Zonesion"命令，解析自定义 FinSH 控制台命令后，调用 MULTIEDIT_AddText()函数在 MULITEDIT 中添加字符，令 LCD 显示相应字符"Zonesion"，如图 5.14 所示。

```
 \ | /
- RT -      Thread Operating System
 / | \      4.1.0 build Nov   4 2022 11:36:35
 2006 - 2022 Copyright by RT-Thread team
msh >guiCtrl show Zonesion
msh >
```

图 5.14　在 LCD 中显示"Zonesion"

（3）在 MobaXterm 串口终端 FinSH 控制台中输入"guiCtrl clear"命令，解析自定义 FinSH 控制台命令后，调用 MULTIEDIT_SetText()函数设置 MULITEDIT 显示为空，清除 LCD 中显示的所有字符，如图 5.15 所示。

```
 \ | /
- RT -      Thread Operating System
 / | \      4.1.0 build Nov   4 2022 11:36:35
 2006 - 2022 Copyright by RT-Thread team
msh >guiCtrl show Zonesion
msh >guiCtrl clear
msh >
```

图 5.15　清除 LCD 中显示的字符

5.5.4　小结

本节主要介绍 emWin 的 GUI 控件管理方式。通过本节的学习，读者可掌握 emWin 的控件管理接口的应用。

第 6 章
RT-Thread 网络应用开发技术

6.1 LWIP 应用开发

LWIP（LightWeight IP）是一种轻量级的开源 TCP/IP 协议栈，专门用于嵌入式系统和小型设备。LWIP 提供了可以在嵌入式设备上实现的网络通信协议，如 IP、TCP、UDP、ICMP和各种应用层协议。LWIP 非常适合资源有限的嵌入式系统，因为它对内存占用和处理器的要求都比较低。

本节的要求如下：

- 了解网卡的原理和概念。
- 熟悉 LWIP 的原理。
- 掌握网卡的管理方式。
- 掌握 RT-Thread 网卡管理接口的应用。

6.1.1 原理分析

6.1.1.1 网卡简介

在 RT-Thread 中，每一个用于网络连接的设备都可以注册成网卡，也称为网络接口设备（Network Interface Device）。为了适配更多的网络接口设备，避免系统对单一网络接口设备的依赖，RT-Thread 提供了 netdev 组件来管理和控制网卡。

netdev 组件的主要作用是解决网络连接问题，统一管理网络接口设备的信息与网络连接状态，提供统一的网卡管理接口。其主要特点如下：

- 抽象网卡概念，每个网络接口设备都可以注册成网卡。
- 可查询多种网络接口设备信息和网络连接状态。
- 建立了网卡列表和默认网卡，可用于网络连接的切换。
- 提供可网卡管理接口，可设置 IP 地址、DNS 服务器地址、网卡状态等。
- 统一管理网卡的调试命令（如 ping、ifconfig、netstat、dns 等命令）。

1）网卡和协议栈

RT-Thread 目前支持三种协议栈，即 LWIP、AT Socket、WIZnet 全硬件 TCP/IP，每种协议栈对应一种协议簇类型（family），上述三种协议栈对应的协议簇类型分别为 AF_INET、AF_AT、AF_WIZ。

网卡的初始化和注册是以协议簇类型为基础进行的，每种网卡对应唯一的协议簇类型。

Socket 的创建是在网卡的基础进行的，Socket 对应唯一的网卡。协议簇类型、网卡和 Socket 之间的关系如图 6.1 所示。

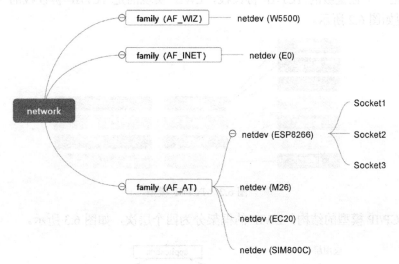

图 6.1　协议簇类型、网卡和 Socket 之间的关系

2）网卡结构体

每个网卡对应唯一的网卡结构体。网卡结构体包含该网卡的主要信息和实时状态，用于系统获取网卡信息和设置网卡。网卡结构体的成员：

```
//网卡结构体
struct netdev
{
    rt_slist_t list;                                        //网卡列表

    char name[RT_NAME_MAX];                                 //网卡名称
    ip_addr_t ip_addr;                                      //IP 地址
    ip_addr_t netmask;                                      //子网掩码地址
    ip_addr_t gw;                                           //网关地址
    ip_addr_t dns_servers[NETDEV_DNS_SERVERS_NUM];          //DNS 服务器地址
    uint8_t hwaddr_len;                                     //硬件地址长度
    uint8_t hwaddr[NETDEV_HWADDR_MAX_LEN];                  //硬件地址

    uint16_t flags;                                         //网卡状态位
    uint16_t mtu;                                           //网卡最大传输单元
    const struct netdev_ops *ops;                           //网卡操作回调函数

    netdev_callback_fn status_callback;                     //网卡状态改变时的回调函数
    netdev_callback_fn addr_callback;                       //网卡地址改变时的回调函数

#ifdef RT_USING_SAL
    void *sal_user_data;                                    //协议簇相关参数
#endif                                                      //RT_USING_SAL
    void *user_data;                                        //预留用户数据
};
```

6.1.1.2　LWIP 简介

LWIP 是一个轻量级的 TCP/IP 协议栈，LWIP 实现的是 TCP/IP 协议栈的一部分功能。TCP/IP 模型如图 6.2 所示。

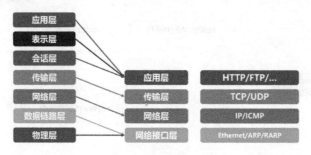

图 6.2　TCP/IP 模型

根据 TCP/IP 模型的结构，LWIP 的框架分为四个层次，如图 6.3 所示。

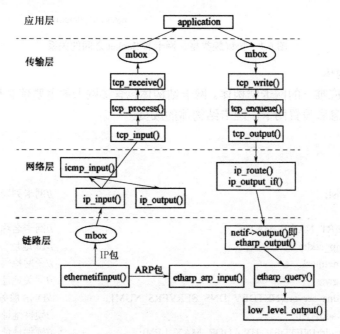

图 6.3　LWIP 的框架

LWIP 的主要特点如下：

⊃ 轻量级设计：LWIP 是轻量级的协议栈，可用于嵌入式系统等资源限制的设备。

⊃ 支持多种协议：LWIP 支持常见的网络协议，如 IPv4、IPv6、TCP、UDP、ICMP 等。

⊃ 裁减能力：用户可以根据应用的需求选择性地启用或禁用 LWIP 的部分功能。

⊃ 异步事件驱动：LWIP 是基于异步事件驱动的协议栈，而不是采用阻塞式的同步模型，这意味着它可以通过事件驱动机制来进行网络操作。

⊃ 支持 TCP/IP 和 UDP：LWIP 实现了 TCP/IP 和 UDP，开发者可以基于这些协议轻松地构建网络应用。

LWIP 常用于网络通信、远程监控、传感器数据采集、IoT 设备、嵌入式 Web 服务器等嵌入式系统，为嵌入式开发人员提供了一个强大的工具。要使用 LWIP，开发人员需要将 LWIP

集成到目标硬件和嵌入式应用程序中，并进行相应的配置。

6.1.1.3　网卡的管理方式

本节采用的是 DM9051 芯片，该芯片是一个具有 SPI 接口的独立以太网控制器，因此对网卡的操作相当于对 SPI 设备的操作。

1）挂载网卡

挂载网卡相当于挂载 SPI 设备，详见 3.9.1.2 节。

2）配置网卡

配置网卡相当于配置 SPI 设备，通过下面的函数可配置网卡：

```
rt_err_t rt_spi_configure(struct rt_spi_device *device, struct rt_spi_configuration *cfg)
```

该函数的详细说明请参考 3.9.1.2 节，该函数会将 cfg 指向的配置参数保存到网卡的控制块中。指针 cfg 是结构体指针，结构体 rt_spi_configuration 的成员如下：

```
struct rt_spi_configuration
{
    rt_uint8_t mode;                    //模式
    rt_uint8_t data_width;              //数据宽度，可取 8 位、16 位、32 位
    rt_uint16_t reserved;               //保留
    rt_uint32_t max_hz;                 //最大频率
};
```

其中的模式包含最高有效位（MSB）在前或最低有效位（LSB）在前、主从模式、时序模式等，模式可设置为多种宏定义的组合，例如：

```
//设置数据传输顺序是 MSB 在前还是 LSB 在前
#define RT_SPI_LSB    (0<<2)                            //bit[2]为 0 表示 LSB 在前
#define RT_SPI_MSB    (1<<2)                            //bit[2]为 1 表示 MSB 在前
//设置 SPI 的主从模式
#define RT_SPI_MASTER    (0<<3)                         //SPI 主机
#define RT_SPI_SLAVE    (1<<3)                          //SPI 从机
//设置时钟极性和时钟相位
#define RT_SPI_MODE_0 (0 | 0)                           //CPOL = 0、CPHA = 0
#define RT_SPI_MODE_1 (0 | RT_SPI_CPHA)                 //CPOL = 0、CPHA = 1
#define RT_SPI_MODE_2 (RT_SPI_CPOL | 0)                 //CPOL = 1、CPHA = 0
#define RT_SPI_MODE_3 (RT_SPI_CPOL | RT_SPI_CPHA)       //CPOL = 1、CPHA = 1
#define RT_SPI_CS_HIGH    (1<<4)                        //CS 信号高电平有效
#define RT_SPI_NO_CS    (1<<5)                          //无 CS 信号
#define RT_SPI_3WIRE    (1<<6)                          //共享 SI/SO 引脚
#define RT_SPI_READY    (1<<7)                          //从机拉低暂停
```

网卡配置示例如下：

```
struct rt_spi_configuration cfg;
cfg.data_width = 8;
cfg.mode = RT_SPI_MASTER | RT_SPI_MODE_0 | RT_SPI_MSB;
cfg.max_hz = 20 * 1000 *1000;
rt_spi_configure(spi_dev, &cfg);
```

3）访问网卡

访问网卡和访问 SPI 设备类似，如果需要在发送数据后接收数据，则可以使用下面的函数：

```
/***********************************************************************
 * 名称：rt_spi_send_then_recv()
 * 功能：发送数据后接收数据
 * 参数：device 表示 SPI 从机句柄；send_buf 表示指向发送数据缓冲区的指针；send_length 表示发送
数据的字节数；recv_buf 表示指向接收数据缓冲区的指针；recv_length 表示接收数据的字节数
 * 返回：RT_EOK 表示成功；RT_EIO 表示失败
 ***********************************************************************/
rt_err_t rt_spi_send_then_recv(struct rt_spi_device*  device, const void*  send_buf, rt_size_t
                            send_length, void*  recv_buf, rt_size_t  recv_length);
```

访问网卡的其他函数如下：

- rt_spi_transfer()：传输一次数据。
- rt_spi_send()：发送一次数据。
- rt_spi_recv()：接收一次数据。
- rt_spi_send_then_send()：连续两次发送数据。

4）查找网卡

查找网卡相当于查找 SPI 设备，详见 3.9.1.3 节。

5）注册网卡

在完成网卡的初始化后，通过下面的函数可将网卡注册到网卡列表中：

```
/***********************************************************************
 * 名称：netdev_register()
 * 功能：注册网卡
 * 参数：netdev 表示网卡对象；name 表示网卡名称；user_data 表示用户使用数据
 * 返回：0 表示网卡注册成功；-1 表示网卡注册失败
 ***********************************************************************/
int netdev_register(struct netdev*  netdev, const char*  name, void*  user_data);
```

上面的函数无须用户调用，系统将在网卡初始化完成后自动调用该函数。

6.1.1.4 网卡的初始化函数

本书使用 dm9051_auto_init()函数来初始化网卡，该函数对 dm9051_probe()函数进行了封装。dm9051_probe()函数的原型如下：

```
/***********************************************************************
 * 名称：dm9051_probe()
 * 功能：DM9051 设备初始化
 * 参数：spi_dev_name 表示网卡对应的 SPI 设备名称；device_name 表示网卡名称；rst_pin 表示复位
引脚；int_pin 表示中断请求引脚
 ***********************************************************************/
int dm9051_probe(const char *spi_dev_name, const char *device_name, int rst_pin, int int_pin)
```

6.1.2　开发设计与实践

6.1.2.1　软件设计

软件设计流程如图 6.4 所示。

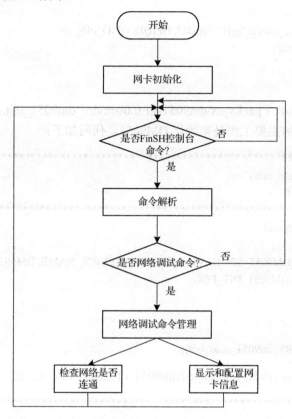

图 6.4　软件设计流程

6.1.2.2　功能设计与核心代码设计

本节首先对网卡进行初始化，然后通过相应的接口使用网卡。

1）主函数（zonesion/app/main.c）

主函数使用 rt_hw_spi_device_attach() 函数将 DM9051（对应的 SPI 设备名称为 spi11）挂载在 SPI1 总线上。由于在 dm9051_auto_init() 函数中将 DM9051 的名称设置为 spi11，因此需要先确定 spi11 的 CS 信号，然后调用 dm9051_auto_init() 函数完成网卡的初始化，在网卡初始化完成后退出主函数。代码如下：

```
/**********************************************************************************
* 名称：main()
* 功能：网卡初始化
**********************************************************************************/
#include <rtthread.h>

#define DBG_TAG "main"
```

```
#define DBG_LVL DBG_LOG
#include <rtdbg.h>

#include "drv_spi.h"
int dm9051_auto_init(void);
int main(void)
{
    rt_hw_spi_device_attach("spi1", "spi11", GPIOD, GPIO_PIN_3);
    dm9051_auto_init();
    return 0;
}
```

2）网卡初始化函数（packages\dm9051-v1.0.0\src\drv_dm9051_init.c）

网卡初始化函数的主要工作是初始化 DM9051。代码如下：

```
/*******************************************************************************
* 名称：dm9051_auto_init()
* 功能：网卡初始化
*******************************************************************************/
int dm9051_auto_init(void)
{
    dm9051_probe(DM9051_SPI_DEVICE, DM9051_DEVICE_NAME, DM9051_RST_PIN,
                DM9051_INT_PIN);

    return 0;
}
//INIT_ENV_EXPORT(dm9051_auto_init);
```

3）dm9051_probe( )函数（packages\dm9051-v1.0.0\src\drv_dm9051.c）

```
/*******************************************************************************
* 名称：dm9051_probe()
* 功能：DM9051 初始化
*******************************************************************************/
int dm9051_probe(const char *spi_dev_name, const char *device_name, int rst_pin, int int_pin)
{
    struct rt_dm9051_eth *eth;
    uint32_t device_id;
    struct rt_spi_device *spi_device;
    spi_device = (struct rt_spi_device *)rt_device_find(spi_dev_name);
    if (spi_device == RT_NULL)
    {
        LOG_E("[%s:%d] spi device %s not found!.", __FUNCTION__, __LINE__, spi_dev_name);
        return -RT_ENOSYS;
    }
    //配置 SPI 设备
    {
        struct rt_spi_configuration cfg;
        cfg.data_width = 8;
```

```
            cfg.mode = RT_SPI_MODE_0 | RT_SPI_MSB;
            //设置 DM9051_SPI_MAX_HZ 的值，默认值为 20 MHz，最大值为 50 MHz
            cfg.max_hz = 20 * 1000 * 1000;
            rt_spi_configure(spi_device, &cfg);
        }
        device_id = DM9051_read_reg(spi_device, DM9051_VIDL);
        device_id |= DM9051_read_reg(spi_device, DM9051_VIDH) << 8;
        device_id |= DM9051_read_reg(spi_device, DM9051_PIDL) << 16;
        device_id |= DM9051_read_reg(spi_device, DM9051_PIDH) << 24;
        LOG_I("[%s L%d] device_id: %08X", __FUNCTION__, __LINE__, device_id);
        if(device_id != DM9051_ID)
        {
            return -1;
        }
        device_id = DM9051_read_reg(spi_device, DM9051_CHIPR);
        LOG_I("[%s L%d] CHIP Revision: %02X", __FUNCTION__, __LINE__, device_id);
        eth = rt_calloc(1, sizeof(struct rt_dm9051_eth));
        if(!eth)
        {
            return -1;
        }
        eth->spi_device = spi_device;
        eth->rst_pin = rst_pin;
        eth->int_pin = int_pin;
        //OUI 00-60-6E Davicom Semiconductor, Inc.
        eth->dev_addr[0] = 0x00;
        eth->dev_addr[1] = 0x60;
        eth->dev_addr[2] = 0x6E;
        eth->dev_addr[3] = 12;
        eth->dev_addr[4] = 34;
        eth->dev_addr[5] = 56;
        rt_pin_mode(eth->int_pin, PIN_MODE_INPUT_PULLDOWN);
        device_id = rt_pin_attach_irq(eth->int_pin, PIN_IRQ_MODE_RISING, dm9051_isr, eth);
        LOG_D("[%s L%d] rt_pin_attach_irq #%d res:%d \n", __FUNCTION__, __LINE__, eth->int_pin,
device_id);
    //init rt-thread device struct
    eth->parent.parent.type = RT_Device_Class_NetIf;
#ifdef RT_USING_DEVICE_OPS
    eth->parent.parent.ops = &dm9051_ops;
#else
    eth->parent.parent.init = dm9051_init;
    eth->parent.parent.open = dm9051_open;
    eth->parent.parent.close = dm9051_close;
    eth->parent.parent.read = dm9051_read;
    eth->parent.parent.write = dm9051_write;
    eth->parent.parent.control = dm9051_control;
#endif //RT_USING_DEVICE_OPS
    //init rt-thread ethernet device struct
```

```
        eth->parent.eth_rx = dm9051_rx;
        eth->parent.eth_tx = dm9051_tx;
        rt_mutex_init(&eth->lock, "dm9051", RT_IPC_FLAG_FIFO);
        rt_timer_init(&eth->timer, "dm9051", dm9051_timeout, eth, RT_TICK_PER_SECOND/2,
                    RT_TIMER_FLAG_PERIODIC | RT_TIMER_FLAG_SOFT_TIMER);
        dm9051_monitor = eth;
        eth_device_init(&eth->parent, "e0");
        return 0;
}
```

6.1.3 开发步骤与验证

6.1.3.1 硬件部署

（1）准备 ZI-ARMEmbed、ARM 仿真器、MiniUSB 线。

（2）完成仿真器、串口和 12 V 电源之间的连接。

（3）使用 MobaXterm 工具创建串口终端，将串口的波特率、数据位、停止位、校验位、流控制分别设置为 115200、8、1、None、None。

（4）通过网线将 ZI-ARMEmbed 和计算机连接到同一个局域网。

6.1.3.2 项目部署

同 2.1.3.2 节。

6.1.3.3 验证效果

（1）关闭 RT-Thread Studio，拔掉仿真器，按下 ZI-ARMEmbed 上的电源按键重新上电后，可通过 DHCP 自动获取 IP 地址。

```
 \ | /
- RT -     Thread Operating System
 / | \     4.1.0 build Nov  8 2022 11:07:25
 2006 - 2022 Copyright by RT-Thread team
lwIP-2.1.2 initialized!
read W25Qxx ID is: 0x16EF
[I/sal.skt] Socket Abstraction Layer initialize success.
[I/[drv.dm9051] ] [dm9051_probe L674] device_id: 90510A46
[I/[drv.dm9051] ] [dm9051_probe L682] CHIP Revision: 01
msh >[I/[drv.dm9051] ] MAC: 00-60-6E-0C-22-38
```

（2）在 MobaXterm 串口终端 FinSH 控制台中输入命令"ifconfig"，可查看网卡的 IP 地址。

```
 \ | /
- RT -     Thread Operating System
 / | \     4.1.0 build Nov  8 2022 11:07:25
 2006 - 2022 Copyright by RT-Thread team
lwIP-2.1.2 initialized!
read W25Qxx ID is: 0x16EF
[I/sal.skt] Socket Abstraction Layer initialize success.
```

```
[I/[drv.dm9051] ] [dm9051_probe L674] device_id: 90510A46
[I/[drv.dm9051] ] [dm9051_probe L682] CHIP Revision: 01
msh >[I/[drv.dm9051] ] MAC: 00-60-6E-0C-22-38
ifconfig
network interface device: e0 (Default)
MTU: 1500
MAC: 00 60 6e 0c 22 38
FLAGS: UP LINK_UP INTERNET_UP DHCP_ENABLE ETHARP BROADCAST IGMP
ip address: 192.168.100.149
gw address: 192.168.100.1
net mask : 255.255.255.0
dns server #0: 192.168.100.1
dns server #1: 0.0.0.0
msh >
```

（3）在 MobaXterm 串口终端 FinSH 控制台中输入命令"ping 计算机 IP 地址"（计算机 IP 地址需要读者去查看实际的地址），可以看见此时 ZI-ARMEmbed 和计算机之间网络是连通的。

```
         \ | /
     - RT -     Thread Operating System
      / | \     4.1.0 build Nov   8 2022 11:07:25
     2006 - 2022 Copyright by RT-Thread team
lwIP-2.1.2 initialized!
read W25Qxx ID is: 0x16EF
[I/sal.skt] Socket Abstraction Layer initialize success.
[I/[drv.dm9051] ] [dm9051_probe L674] device_id: 90510A46
[I/[drv.dm9051] ] [dm9051_probe L682] CHIP Revision: 01
msh >[I/[drv.dm9051] ] MAC: 00-60-6E-0C-22-38
ifconfig
network interface device: e0 (Default)
MTU: 1500
MAC: 00 60 6e 0c 22 38
FLAGS: UP LINK_UP INTERNET_UP DHCP_ENABLE ETHARP BROADCAST IGMP
ip address: 192.168.100.149
gw address: 192.168.100.1
net mask : 255.255.255.0
dns server #0: 192.168.100.1
dns server #1: 0.0.0.0
msh >ping 192.168.100.200
60 bytes from 192.168.100.200 icmp_seq=0 ttl=128 time=139 ms
60 bytes from 192.168.100.200 icmp_seq=1 ttl=128 time=6 ms
60 bytes from 192.168.100.200 icmp_seq=2 ttl=128 time=4 ms
60 bytes from 192.168.100.200 icmp_seq=3 ttl=128 time=1 ms
msh >
```

（4）在 CMD 命令行窗口中输入"ping 192.168.100.149"命令（ZI-ARMEmbed 板卡 IP 地址），可以看见计算机和 ZI-ARMEmbed 已经连接成功。

```
C:\Users\Administrator>ping 192.168.100.149

正在 Ping 192.168.100.149 具有 32 字节的数据:
来自 192.168.100.149 的回复: 字节=32 时间=1ms TTL=255
来自 192.168.100.149 的回复: 字节=32 时间=1ms TTL=255
来自 192.168.100.149 的回复: 字节=32 时间=5ms TTL=255
来自 192.168.100.149 的回复: 字节=32 时间=1ms TTL=255

192.168.100.149 的 Ping 统计信息:
    数据包: 已发送 = 4，已接收 = 4，丢失 = 0 (0% 丢失)，
往返行程的估计时间(以毫秒为单位):
    最短 = 1ms，最长 = 5ms，平均 = 2ms

C:\Users\Administrator>
```

6.1.4　小结

本节主要介绍网卡的基本原理和基本概念、LWIP、网卡的管理方式。通过本节的学习，读者可掌握网卡管理接口的应用。

6.2 AT Socket 协议栈应用开发

AT 命令是一种用于与调制解调器、无线模块、移动设备等进行交互的控制命令。AT 命令以"AT"开头，后跟具体的命令，用于执行各种操作，如配置设备、发送短信、建立数据连接等。

本节的要求如下：

- ➲ 了解 AT 命令的基本原理与概念。
- ➲ 熟悉 AT Socket 协议栈。
- ➲ 掌握 RT-Thread AT Socket 协议栈管理接口的应用。

6.2.1　原理分析

6.2.1.1　AT 命令

AT 命令最初是用于与调制解调器进行通信的，后来被广泛用于与其他串口设备（如嵌入式模块、GPS 接收器等）进行通信。如今，AT 命令成为一种应用于 AT 服务器（AT Server）与 AT 客户端（AT Client）间的设备连接与数据通信的方式。AT 命令通信的基本结构如图 6.6 所示。

图 6.5　AT 命令通信的基本结构

AT 命令由三个部分组成，分别是前缀、主体和结束符，其中前缀是 AT，主体由命令、参数和可能用到的数据组成，结束符为\<CR\>和\<LF\>（回车符和换行符，即\r\n）。AT 命令通信需要由 AT 服务器和 AT 客户端共同完成。AT 服务器主要用于接收 AT 客户端发送的命令，判断接收到的命令及参数格式；发送响应数据或者 URC 数据。AT 客户端主要用于发送命令、等待 AT 服务器响应，并对 AT 服务器发送的响应数据或 URC 数据进行解析处理，从而获取相关信息。AT 服务器和 AT 客户端之间可采用多种数据通信方式（如 UART、SPI 等），目前最常用的是 UART。AT 服务器向 AT 客户端发送的数据分成两种，即响应数据和 URC 数据。响应数据是指 AT 客户端发送命令之后收到的 AT 服务器响应状态和信息。URC 数据是指 AT 服务器主动发送给 AT 客户端的数据，一般出现在一些特殊的情况，如网络连接断开等，这些情况往往需要用户进行相应的操作。

6.2.1.2　AT Socket 协议栈

通过 AT Socket 协议栈，设备既可以作为 AT 客户端通过串口连接其他设备；也可以作为 AT 服务器与其他设备甚至计算机连接，从而发送响应数据或 URC 数据；还可以在本地 Shell 启动 AT CLI 模式，使设备同时支持 AT 服务器和 AT 客户端功能，该模式多用于设备开发调试。

在 RT-Thread 中，AT Socket 协议栈组件的资源占用情况如下：

- ⊃ AT 客户端功能：占用 4.6 KB 的 ROM 和 2 KB 的 RAM。
- ⊃ AT 服务器功能：占用 4 KB 的 ROM 和 2.5 KB 的 RAM。
- ⊃ AT CLI 模式：占用 1.5 KB 的 ROM，几乎不使用 RAM。

AT Socket 协议栈组件占用的资源极小，非常适合资源受限的嵌入式设备。AT Socket 协议栈组件的代码主要位于 rt-thread/components/net/at/目录中。

1）AT 服务器的主要功能

- ⊃ 基础命令：支持多种通用的基础命令（如 ATE、ATZ 等）。
- ⊃ 命令兼容：忽略大小写，提高了命令兼容性。
- ⊃ 命令检测：支持自定义参数表达式，可对接收到的命令参数进行检测。
- ⊃ 命令注册：提供了简单的用户自定义命令添加方式，类似于 finsh/msh 命令添加方式。
- ⊃ 设备调试：提供 AT 服务器模式、AT CLI 模式，主要用于设备调试。

当使用 AT Socket 协议栈组件中的 AT 服务器功能时，需要在 rtconfig.h 中配置 AT 服务器。AT 服务器的配置如表 6.1 所示。

表 6.1　AT 服务器的配置

宏 定 义	描　　述
RT_USING_AT	开启 AT Socket 协议栈组件
AT_USING_SERVER	开启 AT 服务器功能
AT_SERVER_DEVICE	定义 AT 服务器使用的串口通信设备名称，确保设备名称唯一，如 UART3 设备
AT_SERVER_RECV_BUFF_LEN	AT 服务器接收数据的最大长度
AT_CMD_END_MARK_CRLF	判断接收命令的结束符
AT_USING_CLI	开启 AT 服务器命令行交互模式
AT_DEBUG	开启 AT Socket 协议栈组件 DEBUG 模式，可以显示更多调试信息
AT_PRINT_RAW_CMD	实时显示 AT 命令通信数据模式，方便调试

2）AT 客户端的主要功能

- URC 数据处理：完备的 URC 数据处理方式。
- 数据解析：支持自定义响应数据的解析方式，方便获取响应数据中的相关信息。
- 设备调试：提供 AT CLI 模式，主要用于设备调试。
- AT Socket 协议栈：作为 AT 客户端功能的延伸，以 AT 命令收发数据为基础，实现了标准的 BSD Socket API，使用户可以通过 AT 命令完成设备联网和数据通信。
- 支持多客户端：AT Socket 协议栈目前支持多客户端同时运行。

当使用 AT Socket 协议栈组件中的 AT 客户端功能时，需要在 rtconfig.h 中配置 AT 客户端。AT 客户端的配置如表 6.2 所示。

表 6.2　AT 客户端的配置

宏 定 义	描　　述
RT_USING_AT	开启 AT Socket 协议栈组件
AT_USING_CLIENT	开启 AT 客户端功能
AT_CLIENT_NUM_MAX	同时支持的 AT 客户端最大数量
AT_USING_SOCKET	用于 AT 客户端支持标准 BSD Socket API，开启 AT Socket 协议栈组件
AT_USING_CLI	开启或关闭 AT CLI 模式
AT_PRINT_RAW_CMD	开启 AT 命令通信数据的实时显示模式，方便调试

3）结构体 at_response

在 AT Socket 协议栈组件中，结构体 at_response 用于定义一个 AT 命令响应数据的控制块，该结构体如下：

```
struct at_response
{
    char *buf;
    rt_size_t buf_size;
    * == 0: the response data will auto return when received 'OK' or 'ERROR'
    * != 0: the response data will return when received setting lines number data
    rt_size_t line_num;
    rt_size_t line_counts;
    rt_int32_t timeout;
};
typedef struct at_response *at_response_t;
```

其中，buf 用于存放接收到的响应数据，需要注意的是 buf 中存放的响应数据并不是原始的响应数据，而是去除了原始响应数据的结束符（"\r\n"）。buf 中的每行响应数据以 "\0" 分割，方便按行获取相关信息。buf_size 表示用户自定义的接收响应数据的最大长度，由用户根据 AT 命令返回值的长度定义。line_num 表示用户自定义的响应数据的行数，如果对行数没有限制，则可将其设置为 0。line_counts 表示响应数据的总行数。timeout 表示用户自定义的响应数据的最大响应时间。结构体 at_response 中 buf_size、line_num、timeout 为限制条件，需要在创建结构体时设置，其他参数是存放响应数据的参数，用于后面的数据解析。

6.2.1.3 AT Socket 协议栈的管理方式

1）AT 客户端初始化

配置 AT 客户端后，需要在启动时对它进行初始化，并开启 AT 客户端功能。如果程序已经对 AT Socket 协议栈组件进行了初始化，则不需要再额外对 AT 客户端进行初始化，否则需要通过下面的函数初始化 AT 客户端。

```
/*****************************************************************************
 * 名称：at_client_init()
 * 功能：AT 客户端初始化
 * 参数：dev_name 表示 AT 客户端设备名称；recv_bufsz 表示缓冲区可接收数据的最大长度
 * 返回：0 表示初始化成功；-1 表示初始化失败；-5 表示无缓冲区
 *****************************************************************************/
int at_client_init(const char*   dev_name, rt_size_t   recv_bufsz);
```

AT 客户端初始化函数属于应用层函数，需要在使用 AT 客户端功能或者使用 AT CLI 模式前调用。函数 at_client_init()可对 AT 客户端设备、AT 客户端移植函数、AT 客户端使用的信号量和互斥量等资源进行初始化，并创建 at_client 线程，该线程可用于在 AT 客户端中接收并解析数据。

2）创建结构体 at_response

通过下面的函数可创建结构体 at_response：

```
/*****************************************************************************
 * 名称：at_create_resp()
 * 功能：创建结构体 at_response
 * 参数：buf_size 表示用户自定义的接收响应数据的最大长度；line_num 表示用户自定义的响应数据
的行数；timeout 表示用户自定义的响应数据的最大响应时间
 * 返回：非 NULL 表示成功，返回指向结构体 at_response 的指针；NULL 表示失败，内存不足
 *****************************************************************************/
at_response_t at_create_resp(rt_size_t buf_size, rt_size_t line_num, rt_int32_t timeout)
```

3）设置结构体 at_response 的参数

通过下面的函数可设置结构体 at_response 的参数：

```
/*****************************************************************************
 * 名称：at_resp_set_info()
 * 功能：设置响应结构体参数
 * 参数：resp 表示指向结构体 at_response 的指针；buf_size、line_num 和 timeout 同函数 at_create_resp()
 * 返回：非 NULL 表示成功，返回指向结构体 at_response 的指针；NULL 表示失败，内存不足
 *****************************************************************************/
at_response_t at_resp_set_info(at_response_t   resp, rt_size_t   buf_size, rt_size_t   line_num, rt_int32_t
timeout);
```

4）向 AT 服务器发送命令并等待响应

通过下面的函数可向 AT 服务器发送命令并等待响应：

```
/*****************************************************************************
 * 名称：at_obj_exec_cmd()
 * 功能：向 AT 服务器发送命令并等待响应
 * 参数：client 表示当前 AT 客户端对象；resp 表示 AT 响应对象，当不需要响应时使用 RT_NULL；
```

md_expr 表示 AT 命令表达式

　　* 返回：0 表示成功；-1 表示响应状态错误；-2 表示等待超时；-7 表示进入 AT CLI 模式
***/

int at_obj_exec_cmd(at_client_t client, at_response_t resp, const char *cmd_expr, ...)

　　5）解析指定关键字行的响应数据

　　通过下面的函数可解析指定关键字行的响应数据：

/**

　　* 名称：at_resp_parse_line_args_by_kw()

　　* 功能：解析指定关键字行的响应数据

　　* 参数：resp 表示指向结构体 at_response 的指针；keyword 表示关键字；resp_expr 表示自定义的参数解析表达式

　　* 返回：>0 表示成功，返回解析成功的参数个数；0 表示失败，无匹配参数解析表达式的参数；-1 表示失败，参数解析错误
**/

int at_resp_parse_line_args_by_kw(at_response_t resp, const char *keyword, const char *resp_expr, ...);

6.2.2　开发设计与实践

6.2.2.1　软件设计

软件设计流程如图 6.6 所示。

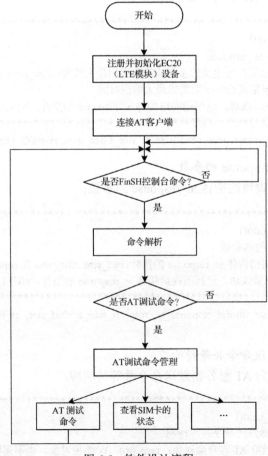

图 6.6　软件设计流程

6.2.2.2　功能设计与核心代码设计

1）主函数（zonesion/app/main.c）

主函数的主要工作是注册并初始化 EC20（LTE 模块）设备，完成之后退出主函数。代码如下：

```
/*******************************************************************************
* 名称：main()
* 功能：注册 EC20（LTE 模块）设备注册
*******************************************************************************/
int ec20_device_register(void);
int main(void)
{
    //AT 客户端初始化已经完成，此处不再演示
    ec20_device_register();                              //注册并初始化 EC20（LTE 模块）设备
    return 0;
}
```

2）EC20 设备初始化函数（packages\at_device-latest\class\ec20\at_device_ec20.c）

EC20 设备初始化函数的主要工作是初始化 AT 客户端、将 EC20 设备添加到网络设备列表、初始化 EC20 设备的引脚、初始化 EC20 设备网络。代码如下：

```
/*******************************************************************************
* 名称：ec20_init()
* 功能：EC20 设备初始化函数
* 参数：at_device 表示 EC20 设备句柄
* 返回：RT_EOK 表示初始化成功；_RT_ERROR 表示初始化失败
*******************************************************************************/
static int ec20_init(struct at_device *device)
{
    struct at_device_ec20 *ec20 = (struct at_device_ec20 *) device->user_data;
    at_client_init(ec20->client_name, ec20->recv_line_num);    //初始化 AT 客户端
    device->client = at_client_get(ec20->client_name);
    if (device->client == RT_NULL)
    {
        LOG_E("get AT client(%s) failed.", ec20->client_name);
        return -RT_ERROR;
    }
    //register URC data execution function
#ifdef AT_USING_SOCKET
    ec20_socket_init(device);
#endif
    device->netdev = ec20_netdev_add(ec20->device_name);       //将 EC20 设备添加到网络设备列表
    if (device->netdev == RT_NULL)
    {
        LOG_E("add netdev(%s) failed.", ec20->device_name);
        return -RT_ERROR;
    }
    if (ec20->power_pin != -1 && ec20->power_status_pin != -1)  //初始化 EC20 设备引脚
```

```
    {
        rt_pin_mode(ec20->power_pin, PIN_MODE_OUTPUT);
        rt_pin_mode(ec20->power_status_pin, PIN_MODE_INPUT);
    }
    return ec20_netdev_set_up(device->netdev);                    //初始化 EC20 设备网络
}
```

3）EC20 设备初始化线程入口函数（packages\at_device-latest\class\ec20\at_device_ec20.c）

EC20 设备初始化线程入口函数的其主要工作是为 EC20 设备上电、启动 EC20 设备、注册相应的 AT 命令等。代码如下：

```
/****************************************************************************
 * 名称：ec20_init_thread_entry (void *parameter)
 * 功能：EC20 设备初始化线程入口函数
 * 参数：parameter 表示入口函数参数，在线程创建时传入
 ****************************************************************************/
/*initialize for ec20
static void ec20_init_thread_entry(void *parameter)
{
#define INIT_RETRY        5
#define CIMI_RETRY        10
#define CSQ_RETRY         20
#define CREG_RETRY        10
#define CGREG_RETRY       20
int i, qi_arg[3] = {0};
int retry_num = INIT_RETRY;
char parsed_data[20] = {0};
rt_err_t result = RT_EOK;
at_response_t resp = RT_NULL;
struct at_device *device = (struct at_device *) parameter;
struct at_client *client = device->client;
resp = at_create_resp(128, 0, rt_tick_from_millisecond(300));
if (resp == RT_NULL)
{
    LOG_E("no memory for resp create.");
    return;
}
LOG_D("start init %s device.", device->name);
while (retry_num--)
{
    ec20_power_on(device);                                  //为 EC20 设备上电
    rt_thread_mdelay(1000);
    //等待 EC20 设备启动完成，每 500 ms 发送一次 AT 命令
    if (at_client_obj_wait_connect(client, EC20_WAIT_CONNECT_TIME))
    {
        result = -RT_ETIMEOUT;
        goto __exit;
    }
```

```
    AT_SEND_CMD(client, resp, 0, 300, "ATV1");              //设置响应格式为 ATV1
    AT_SEND_CMD(client, resp, 0, 300, "ATE0");              //失能响应
    //Use AT+CMEE=2 to enable result code and use verbose values
    AT_SEND_CMD(client, resp, 0, 300, "AT+CMEE=2");
    AT_SEND_CMD(client, resp, 0, 300, "AT+IPR?");           //获取波特率命令
    at_resp_parse_line_args_by_kw(resp, "+IPR:", "+IPR: %d", &i);
    LOG_D("%s device baudrate %d", device->name, i);
    AT_SEND_CMD(client, resp, 0, 300, "ATI");               //获得模块版本号命令
    for (i = 0; i < (int) resp->line_counts - 1; i++)       //显示模块版本号
    {
        LOG_D("%s", at_resp_get_line(resp, i + 1));
    }
    AT_SEND_CMD(client, resp, 0, 300, "AT+GSN");            //使用 AT+GSN 命令查询模块的 IMEI
    AT_SEND_CMD(client, resp, 2, 5 * 1000, "AT+CPIN?");     //检查 SIM 卡命令
    if (!at_resp_get_line_by_kw(resp, "READY"))
    {
        LOG_E("%s device SIM card detection failed.", device->name);
        result = -RT_ERROR;
        goto __exit;
    }
    //waiting for dirty data to be digested
    rt_thread_mdelay(10);
}
```

6.2.3　开发步骤与验证

6.2.3.1　硬件部署

（1）准备 ZI-ARMEmbed、ARM 仿真器、MiniUSB 线。

（2）完成仿真器、串口和 12 V 电源之间的连接。

（3）使用 MobaXterm 工具创建串口终端，将串口的波特率、数据位、停止位、校验位、流控制分别设置为 115200、8、1、None、None。

（4）EC20 设备的安装。在安装 EC20 设备时要注意插入插槽要牢固可靠，在插入 EC20 设备时应尽可能保证螺钉固定正确。在固定 EC20 设备后，还需要连接 EC20 设备的天线，如图 6.7 所示，将天线连接到圆框内的天线柱。

图 6.7　连接 EC20 设备的天线

（5）4G 卡和 4G 天线的安装。4G 天线的接口位置和 4G 卡的安装位置如图 6.8 的方框和圆框所示。

图 6.8   4G 天线的接口位置和 4G 卡的安装位置

（6）在 EC20 设备上电后，当 4G 模块连接网络正常时，ZI-ARMEmbed 中的 LED［见图 6.9 的圆框中的 LED（绿色）］先点亮 2 s 后熄灭一次，否则该 LED 熄灭 2 s 后点亮一次。

图 6.9   4G 模块连接正常的指示灯

6.2.3.2   项目部署

同 2.1.3.2 节。

6.2.3.3   验证效果

（1）关闭 RT-Thread Studio，拔掉仿真器，按下 ZI-ARMEmbed 上的电源按键重新上电。

（2）在 MobaXterm 串口终端 FinSH 控制台中输入 "list_device" 命令，可显示系统中所有的设备信息，包括设备名称、设备类型和设备被打开次数。

```
 \ | /
- RT -     Thread Operating System
 / | \     4.1.0 build Nov  9 2022 15:30:58
 2006 - 2022 Copyright by RT-Thread team
[I/sal.skt] Socket Abstraction Layer initialize success.
[I/at.clnt] AT client(V1.3.1) on device uart4 initialize success.
msh >[I/at.dev.ec20] e0 device network operator: CHINA MOBILE
[I/at.dev.ec20] e0 device IP address: 10.87.16.143
[I/at.dev.ec20] e0 device network initialize success.
msh >list_device
device          type              ref count
--------  --------------------  ----------
uart4      Character Device        1
uart1      Character Device        2
```

```
pin        Miscellaneous Device 0
msh >
```

（3）在 MobaXterm 串口终端 FinSH 控制台中输入命令"at client uart4"，可执行 AT 调试命令，FinSH 控制台输出"Welcome to using RT-Thread AT command client cli"。

```
 \ | /
- RT -       Thread Operating System
 / | \       4.1.0 build Nov   9 2022 15:30:58
 2006 - 2022 Copyright by RT-Thread team
[I/sal.skt] Socket Abstraction Layer initialize success.
[I/at.clnt] AT client(V1.3.1) on device uart4 initialize success.
msh >[I/at.dev.ec20] e0 device network operator: CHINA MOBILE
[I/at.dev.ec20] e0 device IP address: 10.87.16.143
[I/at.dev.ec20] e0 device network initialize success.
msh >list_device
device          type         ref count
-------- -------------------- ----------
uart4    Character Device      1
uart1    Character Device      2
pin         Miscellaneous Device 0
msh >at client uart4
========= Welcome to using RT-Thread AT command client cli =========
Cli will forward your command to server port(uart4). Press 'ESC' to exit.
```

（4）在 MobaXterm 串口终端 FinSH 控制台中输入命令"AT"，可执行 AT 测试命令，FinSH 控制台输出"OK"。

```
 \ | /
- RT -       Thread Operating System
 / | \       4.1.0 build Nov   9 2022 15:30:58
 2006 - 2022 Copyright by RT-Thread team
[I/sal.skt] Socket Abstraction Layer initialize success.
[I/at.clnt] AT client(V1.3.1) on device uart4 initialize success.
msh >[I/at.dev.ec20] e0 device network operator: CHINA MOBILE
[I/at.dev.ec20] e0 device IP address: 10.87.16.143
[I/at.dev.ec20] e0 device network initialize success.
msh >list_device
device          type         ref count
-------- -------------------- ----------
uart4    Character Device      1
uart1    Character Device      2
pin         Miscellaneous Device 0
msh >at client uart4
========= Welcome to using RT-Thread AT command client cli =========
Cli will forward your command to server port(uart4). Press 'ESC' to exit.
msh >AT

OK
```

（5）在 MobaXterm 串口终端 FinSH 控制台中输入"AT+CPIN?"命令，可查询 EC20 设备是否检测到手机卡，FinSH 控制台输出"+CPIN: READY"。

```
  \|/
- RT -       Thread Operating System
 /|\      4.1.0 build Nov   9 2022 15:30:58
 2006 - 2022 Copyright by RT-Thread team
[I/sal.skt] Socket Abstraction Layer initialize success.
[I/at.clnt] AT client(V1.3.1) on device uart4 initialize success.
msh >[I/at.dev.ec20] e0 device network operator: CHINA MOBILE
[I/at.dev.ec20] e0 device IP address: 10.87.16.143
[I/at.dev.ec20] e0 device network initialize success.
msh >list_device
device         type              ref count
-------- -------------------- ----------
uart4     Character Device       1
uart1     Character Device       2
pin        Miscellaneous Device 0
msh >at client uart4
======== Welcome to using RT-Thread AT command client cli ========
Cli will forward your command to server port(uart4). Press 'ESC' to exit.
msh >AT

OK
msh >AT+CPIN?

+CPIN: READY

OK
```

（6）在 MobaXterm 串口终端 FinSH 控制台中输入"Esc"可退出系统。

```
  \|/
- RT -       Thread Operating System
 /|\      4.1.0 build Nov   9 2022 15:30:58
 2006 - 2022 Copyright by RT-Thread team
[I/sal.skt] Socket Abstraction Layer initialize success.
[I/at.clnt] AT client(V1.3.1) on device uart4 initialize success.
msh >[I/at.dev.ec20] e0 device network operator: CHINA MOBILE
[I/at.dev.ec20] e0 device IP address: 10.87.16.143
[I/at.dev.ec20] e0 device network initialize success.
msh >list_device
device         type              ref count
-------- -------------------- ----------
uart4     Character Device       1
uart1     Character Device       2
pin        Miscellaneous Device 0
msh >at client uart4
======== Welcome to using RT-Thread AT command client cli ========
```

Cli will forward your command to server port(uart4). Press 'ESC' to exit.
msh >**AT**

OK
msh >**AT+CPIN?**

+CPIN: READY

OK
msh >**Esc**

6.2.4　小结

本节主要介绍 AT Socket 协议栈的基本原理和基本概念。通过本节的学习，读者可掌握 AT Socket 协议栈管理接口的应用。

6.3 MQTT 协议应用开发

消息队列遥测传输（Message Queuing Telemetry Transport，MQTT）协议是一种轻量级的、基于发布–订阅模式的通信协议，专门用于低带宽、不稳定或资源受限的网络环境中的物联网（IoT）和机器到机器（M2M）通信。MQTT 协议最初由 IBM 开发，现已成为一个开放的 OASIS 标准。MQTT 协议具有可伸缩性，支持可靠的消息传输，可用于不同的网络环境，在物联网、移动互联网、智能硬件、车联网、电力能源等领域得到了广泛应用。

本节的要求如下：

- 了解 MQTT 协议的基本概念与工作原理。
- 掌握 MQTT 协议的管理方式。
- 掌握 RT-Thread MQTT 协议管理接口的应用。

6.3.1　原理分析

6.3.1.1　MQTT 协议概述与组成

1）MQTT 协议的主要特点

（1）采用发布–订阅（Publish-Subscribe）模式：MQTT 协议采用发布–订阅模式，其中的客户端可以订阅感兴趣的主题（Topic）并接收与这些主题相关的消息；发布者将消息发布到特定主题，而订阅该主题的客户端将接收到这些消息。

（2）QoS 级别（Quality of Service）：MQTT 协议提供三个不同的 QoS 级别，用于确保消息传输的可靠性。这些级别包括 0（最多一次交付，无确认）、1（至少一次交付，有确认）和 2（恰好一次交付，有确认）。

（3）保留消息（Retained Messages）：发布者可以选择发布一个主题的保留消息，这意味着订阅该主题的客户端会立即收到最新的消息。

（4）遗嘱消息（Last Will and Testament）：客户端可以指定在其连接断开时发布的消息，这意味着其他客户端可以知道该客户端的状态。

（5）主题层次结构：MQTT 协议的主题是通过层次结构来组织的，类似于文件系统路径。主题可以包含多级子主题，以便组织消息。

（6）低带宽和资源要求：MQTT 协议是轻量级的，适用于资源受限的网络环境和设备，其开销非常小。

（7）持久性会话：MQTT 协议允许客户端创建持久性会话，以便在重新连接时接收之前未接收的消息。

（8）SSL/TLS 支持：MQTT 协议可以通过加密协议（如 SSL/TLS）来提供安全的通信。

2）客户端

一个使用 MQTT 协议的应用程序或者设备总是建立到服务器的网络连接。客户端可以：

（1）与服务器建立连接。

（2）发布其他客户端可能会订阅的消息。

（3）接收其他客户端发布的消息。

（4）退订已订阅的消息。

3）服务器

服务器也称为消息代理（Broker），可以是一个应用程序或一台设备，位于消息发布者和订阅者之间。服务器可以：

（1）接收来自客户的网络连接。

（2）接收发布者发布的消息。

（3）处理来自客户端的订阅请求和退订请求。

（4）向订阅者转发应用程序消息。

4）MQTT 协议中的方法

MQTT 协议中定义了一些方法（也被称为动作），用来表示对指定资源的操作。这里的资源可以是预先存在的数据或动态生成的数据，取决于服务器的实现。

（1）Connect：等待与服务器建立连接。

（2）Disconnect：等待客户端完成所做的工作，并与服务器断开 TCP/IP 会话。

（3）Subscribe：等待完成订阅。

（4）UnSubscribe：等待服务器取消客户端的一个或多个主题订阅。

（5）Publish：客户端发送消息请求，发送完成后返回应用程序。

5）MQTT 协议中的基本概念

（1）订阅（Subscription）：订阅包含主题筛选器和服务质量。订阅会与一个会话（Session）关联，一个会话可以包含多个订阅，每个会话中的每个订阅都有一个不同的主题筛选器。

（2）会话（Session）：客户端与服务器建立的连接就是一个会话，会话可以存在于一个网络内，也可以存在于客户端和服务器之间的多个连接的网络。

（3）主题名（Topic Name）：连接到一个应用程序消息的标签，该标签与服务器的订阅相匹配。服务器会将消息发送给订阅主题的每个客户端。

（4）主题筛选器（Topic Filter）：对主题名中的通配符进行筛选，在订阅表达式中使用，表示订阅所匹配到的多个主题。

（5）负载（Payload）：消息订阅者所接收到的具体内容。

（6）应用消息（Application Message）：应用消息通过 MQTT 协议传输时，有关联的服务质量和主题。

（7）控制报文（Control Packet）：通过网络连接发送的消息数据包，MQTT 协议定义了14 种不同类型的控制报文，其中的 PUBLISH 报文用于传输应用消息。

6.3.1.2　MQTT 协议的工作原理

MQTT 协议的工作原理如图 6.10 所示，其中的发布者和订阅者都是客户端，消息代理是服务器。

图 6.10　MQTT 协议的工作原理

MQTT 协议的使用过程一般遵循以下流程：

（1）发布者通过服务器向指定的主题发布消息。

（2）订阅者通过服务器订阅所需的主题。

（3）订阅成功后，如果发布者向订阅者订阅的主题发布消息，那么订阅者就会收到服务器推送的消息，通过这种方式可以进行高效的数据交换。

6.3.1.3　MQTT 协议的控制报文格式

MQTT 协议通过交换预定义的控制报文来通信。控制报文由三部分组成，如表 6.3 所示。

表 6.3　控制报文的结构

控制报文的组成部分	是否必须包含
固定报头（Fixed Header）	所有的控制报文都包含
可变报头（Variable Header）	部分控制报文包含
负载（Payload）	部分控制报文包含

1）固定报头

所有的控制报文都包含固定报头，其格式如表 6.4 所示。

表 6.4　固定报头的格式

字节的位	7	6	5	4	3	2	1	0
字节 1	控制报文的类型				控制报文类型的标志位			
字节 2	剩余长度							

2）控制报文的类型

控制报文的类型由字节 1 的第 4 位到第 7 位决定，这 4 位的值决定了控制报文的类型，如表 6.5 所示。

表6.5 控制报文的类型

名 字	值	报文传输方向	描 述
Reserved	0	禁止	保留
CONNECT	1	客户端到服务器	客户端请求连接服务器
CONNACK	2	服务器到客户端	连接报文确认
PUBLISH	3	两个方向都允许	发布消息
PUBACK	4	两个方向都允许	QoS 1 消息发布者收到确认
PUBREC	5	两个方向都允许	发布收到（保证交付第一步）
PUBREL	6	两个方向都允许	发布释放（保证交付第二步）
PUBCOMP	7	两个方向都允许	QoS 2 消息发布完成（保证交付第三步）
SUBSCRIBE	8	客户端到服务器	客户端订阅请求
SUBACK	9	服务器到客户端	订阅请求报文确认
UNSUBSCRIBE	10	客户端到服务器	客户端取消订阅请求
UNSUBACK	11	服务器到客户端	取消订阅报文确认
PINGREQ	12	客户端到服务器	心跳请求
PINGRESP	13	服务器到客户端	心跳响应
DISCONNECT	14	两个方向都允许	断开连接通知
AUTH	15	两个方向都允许	认证消息交换

6.3.1.4 MQTT 协议的管理方式

1）申请 MQTT 客户端

通过下面的函数可申请 MQTT 客户端，该函数不仅通过动态申请内存的方式申请了一个 MQTT 客户端结构 mqtt_client_t，还通过 mqtt_init() 函数对 MQTT 客户端进行了初始化，如申请网络组件的内存空间、初始化相关的互斥量、链表等。

```
/********************************************************************************
 * 名称：mqtt_lease()
 * 功能：申请 MQTT 客户端
 * 返回：NULL 表示申请失败；指针 mqtt_client_t 表示申请成功
 ********************************************************************************/
mqtt_client_t *mqtt_lease(void)
```

2）连接 MQTT 服务器

通过下面的函数可连接 MQTT 服务器，该函数的参数只有一个，即 mqtt_client_t 类型的指针。在连接 MQTT 服务器时使用的是非异步的方式，因此必须等待连接到服务器后才能进行下一步的操作。

```
/********************************************************************************
 * 名称：mqtt_connect()
 * 功能：连接 MQTT 服务器
 * 参数：c mqtt_client_t 类型的指针
 * 返回：正数表示连接成功；负数表示连接失败
 ********************************************************************************/
int mqtt_connect(mqtt_client_t* c)
```

3）MQTT 客户端订阅消息

通过下面的函数可以让客户端订阅消息，该函数是采用异步方式订阅报文的，可在主题中使用通配符，如"#""+"等。

```
/****************************************************************************
 * 名称：mqtt_subscribe()
 * 功能：使用 MQTT 客户端订阅一个消息
 * 参数：c 是 mqtt_client_t 类型指针，表示客户端的描述指针；topic_filter 表示订阅的主题；qos 表示
主题的服务质量；handler 表示处理报文的函数，如果未指定则采用默认的处理函数
 * 返回：正数表示订阅消息成功；负数表示订阅消息失败
 ****************************************************************************/
int mqtt_subscribe(mqtt_client_t* c, const char* topic_filter, mqtt_qos_t qos, message_handler_t handler)
```

4）MQTT 客户端发布消息

通过下面的函数可以让客户端发布消息：

```
/****************************************************************************
 * 名称：　mqtt_publish()
 * 功能：使用 MQTT 客户端发布一个消息
 * 参数：c 是 mqtt_client_t 类型指针；topic_filter 表示订阅的主题；msg 表示待发布的消息
 * 返回：正数表示发布消息成功；负数表示发布消息失败
 ****************************************************************************/
int mqtt_publish(mqtt_client_t* c, const char* topic_filter, mqtt_message_t* msg)
```

6.3.2　开发设计与实践

6.3.2.1　软件设计

软件设计流程如图 6.11 所示。

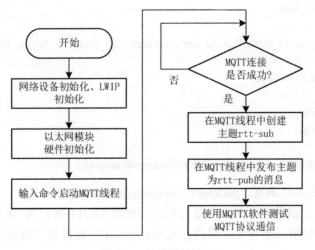

图 6.11　软件设计流程

6.3.2.2　功能设计与核心代码设计

1）主函数（zonesion/app/main.c）

主函数的主要工作是初始化以太网模块硬件，完成初始化之后退出主函数。代码如下：

```
/**********************************************************************
 * 名称：main()
 * 功能：以太网模块硬件初始化
 **********************************************************************/
#include <rtthread.h>

#define DBG_TAG "main"
#define DBG_LVL DBG_LOG
#include <rtdbg.h>

#include "drv_spi.h"
int dm9051_auto_init(void);
int main(void)
{
    //网络设备初始化、LWIP 初始化已经在系统初始化时自动完成
    rt_hw_spi_device_attach("spi1", "spi11", GPIOD, GPIO_PIN_3);
    dm9051_auto_init();                    //以太网模块硬件初始化
    return 0;
}
```

2）MQTT 线程启动函数（packages\kawaii-mqtt-latest\test\test.c）

MQTT 线程启动函数的主要工作是完成 MQTT 线程的初始化和启动。代码如下：

```
/**********************************************************************
 * 名称：ka_mqtt()
 * 功能： MQTT 线程的初始化和启动
 **********************************************************************/
int ka_mqtt(void)
{
    rt_thread_t tid_mqtt;

    tid_mqtt = rt_thread_create("kawaii_demo", kawaii_mqtt_demo, RT_NULL, 2048, 17, 10);
    if (tid_mqtt == RT_NULL) {
        return -RT_ERROR;
    }
    rt_thread_startup(tid_mqtt);
    return RT_EOK;
}
MSH_CMD_EXPORT(ka_mqtt, Kawaii MQTT client test program);
```

3）MQTT 线程入口函数（packages\kawaii-mqtt-latest\test\test.c）

MQTT 线程入口函数的主要工作是先创建和配置 MQTT 客户端，再连接 MQTT 服务器，连接成功后创建指定的订阅消息，最后在 while 循环中发布指定主题的消息。定义如下：

```
/**********************************************************************
 * 名称：kawaii_mqtt_demo()
 * 功能：MQTT 线程入口函数
 * 参数：parameter 为 MQTT 线程入口函数参数
 **********************************************************************/
```

```
static void kawaii_mqtt_demo(void *parameter)
{
    mqtt_client_t *client = NULL;
    rt_thread_delay(6000);
    mqtt_log_init();
    client = mqtt_lease();
    rt_snprintf(cid, sizeof(cid), "rtthread%d", rt_tick_get());
    mqtt_set_host(client, "192.168.100.132");            //设置 URL
    mqtt_set_port(client, "1883");                       //设置端口
    mqtt_set_user_name(client, "rt-thread");             //设置用户名
    mqtt_set_password(client, "rt-thread");              //设置密码
    mqtt_set_client_id(client, cid);
    mqtt_set_clean_session(client, 1);
    KAWAII_MQTT_LOG_I("The ID of the Kawaii client is: %s ",cid);
    mqtt_connect(client);                                //连接 MQTT 服务器
    mqtt_subscribe(client, "rtt-sub", QOS0, sub_topic_handle1);  //订阅消息，主题为 rtt-sub
    while (1) {
        mqtt_publish_handle1(client);                    //发布消息，主题为 rtt-pub
        mqtt_sleep_ms(4 * 1000);                         //延时
    }
}
```

6.3.3　开发步骤与验证

6.3.3.1　建立 MQTT 服务器

（1）安装 Mosquitto，如图 6.12 所示。双击安装包 mosquitto-2.0.15-install-windows-x64.exe，可打开安装向导"Choose Components"，在该向导中选择默认设置后单击"Next"软件，可打开安装向导"Choose Install Location"；在安装向导"Choose Install Location"中选择安装路径后单击"Install"按钮即可安装 Mosquitto。

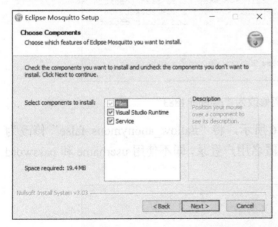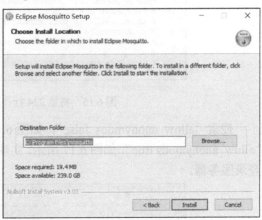

图 6.12　安装 Mosquitto

（2）完成 Mosquitto 的安装后，设置 Path 的环境变量，如图 6.13 所示，添加 Mosquitto 的安装路径。

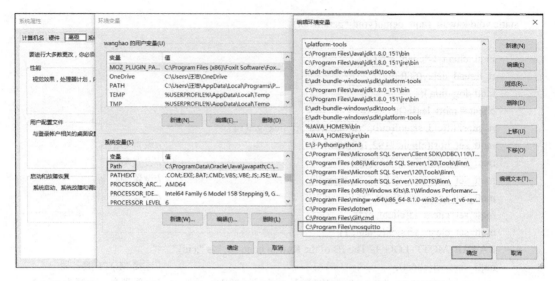

图 6.13   设置 Path 的环境变量

（3）配置服务器文件 mosquitto.conf（C:\Program Files\mosquitto\mosquitto.conf），搜索"listener port-number [ip address/host name/unix socket path]"，会找到图 6.14 所示的代码段，将第 234 行开始处的"#"删除。

```
 mosquitto.conf
227   # On systems that support Unix Domain Sockets, it is also possible
228   # to create a # Unix socket rather than opening a TCP socket. In
229   # this case, the port number should be set to 0 and a unix socket
230   # path must be provided, e.g.
231   # listener 0 /tmp/mosquitto.sock
232   #
233   # listener port-number [ip address/host name/unix socket path]
234   #listener
```

图 6.14   删除第 234 行开始处的"#"

将第 234 行代码修改为"listener 1883"，表示监听端口 1883，如图 6.15 所示。

```
 mosquitto.conf
227   # On systems that support Unix Domain Sockets, it is also possible
228   # to create a # Unix socket rather than opening a TCP socket. In
229   # this case, the port number should be set to 0 and a unix socket
230   # path must be provided, e.g.
231   # listener 0 /tmp/mosquitto.sock
232   #
233   # listener port-number [ip address/host name/unix socket path]
234   listener 1883
```

图 6.15   将第 234 行代码修改为"listener 1883"

搜索"allow_anonymous false"，如图 6.16 所示，将"#allow_anonymous false"修改为"allow_anonymous true"，如图 6.17 所示，实现匿名用户登录，即不使用 username 和 password 登录服务器。

```
524   # Boolean value that determines whether clients that connect
525   # without providing a username are allowed to connect. If set to
526   # false then a password file should be created (see the
527   # password_file option) to control authenticated client access.
528   #
529   # Defaults to false, unless there are no listeners defined in the configuration
530   # file, in which case it is set to true, but connections are only allowed from
531   # the local machine.
532   #allow_anonymous false
533
```

图 6.16   搜索"allow_anonymous false"

```
524   # Boolean value that determines whether clients that connect
525   # without providing a username are allowed to connect. If set to
526   # false then a password file should be created (see the
527   # password_file option) to control authenticated client access.
528   #
529   # Defaults to false, unless there are no listeners defined in the configuration
530   # file, in which case it is set to true, but connections are only allowed from
531   # the local machine.
532   allow_anonymous true
```

图 6.17 将 "allow_anonymous false" 修改为 "allow_anonymous true"

（4）在 Windows 系统中以管理员身份打开命令行窗口，执行命令 "mosquitto install" 即可安装 mosquitto 服务。当再次安装 Mosquitto 时，会出现 "指定的服务已存在" 的错误提示。

```
C:\Users\Administrator>mosquitto install

C:\Users\Administrator>mosquitto install
Error: 指定的服务已存在
```

在 Windows 系统中以管理员身份运行 mosquitto 服务。在服务的列表中找到 Mosquitto Broker，双击这个服务后单击 "启动" 按钮可启动 mosquitto 服务，单击 "启动" 按钮之后，服务状态为 "正在运行" 即可，修改完成后单击 "确认" 按钮，如图 6.18 到图 6.20 所示。如果弹出 "错误 2：系统找不到指定的文件"，则需要先关闭防火墙。

图 6.18 找到服务程序

图 6.19 找到 Mosquitto Broker

图 6.20 mosquitto 服务正在运行

6.3.3.2　建立 MQTT 客户端

免费的 MQTT 客户端软件很多，本节使用的是 MQTTX 软件，其安装包为 MQTTX-Setup-1.8.2-x64.exe。

（1）双击 MQTTX-Setup-1.8.2-x64.exe 后，即可按照自己的设置来安装 MQTTX，如图 6.21 所示。

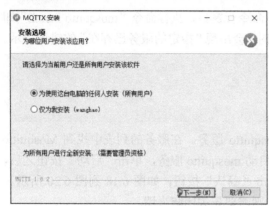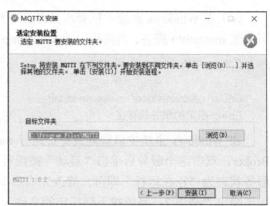

图 6.21　安装 MQTTX

（2）在 MQTTX 软件的运行界面单击"⚙"按钮可打开设置菜单，将"语言"设置为"简体中文"，如图 6.22 所示。

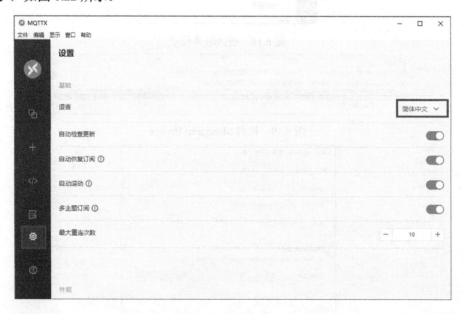

图 6.22　将"语言"设置为"简体中文"

（3）按照图 6.23 配置 MQTTX。

（4）单击"连接"按钮后，如果提示"已连接"，如图 6.24 所示，则表示服务器和客户端设置正常。

（5）单击"添加订阅"按钮，如图 6.25 所示，可弹出"添加订阅"对话框。

在"Topic"中输入"rtt-pub"，如图 6.26 所示，单击"确定"按钮。

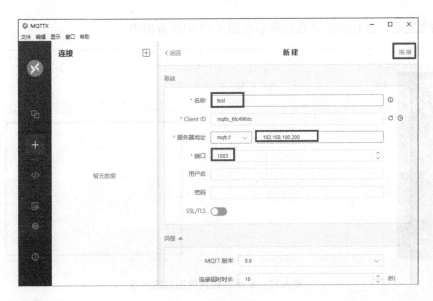

图 6.23　配置 MQTTX

图 6.24　"已连接"状态

图 6.25　单击"添加订阅"按钮

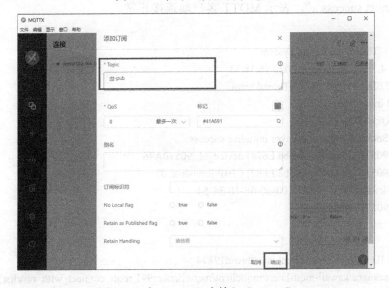

图 6.26　在"Topic"中输入"rtt-pub"

添加主题后，接收到的消息会显示在图 6.27 中的方框中。

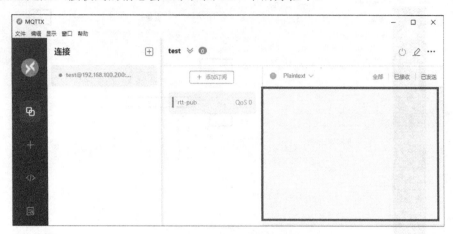

图 6.27   显示接收到的消息的区域

6.3.3.3   项目部署

（1）将工程文件夹（05-Thread）复制到"RT-ThreadStudio/workspace"目录下。

（2）部署公共文件（02-软件资料/01-操作系统/rtt-common.zip）：将 rt-thread 文件夹复制到工程的根目录、将 zonesion/common 文件夹复制到工程的 zonesion 目录下。

（3）修改连接 MQTT 服务器的 IP 地址，在"35-MQTT\packages\kawaii-mqtt-latest\test\test.c"的 static void kawaii_mqtt_demo(void *parameter)函数中，将"mqtt_set_host(client, "192.168.100.193");"中的"192.168.100.193"修改为本机 IP 地址。

（4）使用 RT-Thread Studio 导入本项目软件包。

6.3.3.4   验证效果

（1）关闭 RT-Thread Studio，拔掉仿真器，按下 ZI-ARMEmbed 上的电源按键重新上电。

（2）在 MobaXterm 串口终端 FinSH 控制台中输入"ka_mqtt"命令，等待数秒后，可出现"mqtt connect success..."，表示 MQTT 客户端连接正常。

```
 \ | /
- RT -      Thread Operating System
 / | \      4.1.0 build Nov 10 2022 16:18:13
 2006 - 2022 Copyright by RT-Thread team
lwIP-2.1.2 initialized!
read W25Qxx ID is: 0x16EF
[I/sal.skt] Socket Abstraction Layer initialize success.
[I/[drv.dm9051] ] [dm9051_probe L674] device_id: 90510A46
[I/[drv.dm9051] ] [dm9051_probe L682] CHIP Revision: 01
msh />[I/[drv.dm9051] ] MAC: 00-60-6E-10-34-54
[I/[drv.dm9051] ] link up!
ka_mqtt
msh />
[I] >> The ID of the Kawaii client is: rtthread19874
[I] >> ../packages/kawaii-mqtt-latest/mqttclient/mqttclient.c:991 mqtt_connect_with_results()... mqtt connect
success...
```

此时 MQTTX 软件能够正常接收 ZI-ARMEmbed 发送的消息，如图 6.28 所示。

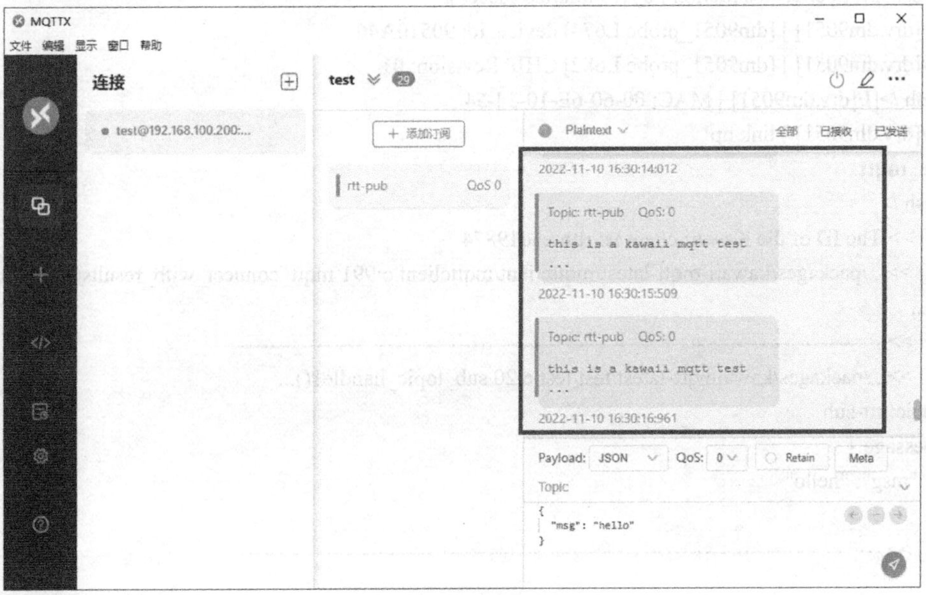

图 6.28　验证效果（一）

（3）在 MQTTX 软件中输入 "rtt-sub"，单击 "●" 按钮后可以在 ZI-ARMEmbed 接收指定消息，如图 6.29 所示。

图 6.29　验证效果（二）

```
 \|/
- RT -     Thread Operating System
 /|\       4.1.0 build Nov 10 2022 16:18:13
 2006 - 2022 Copyright by RT-Thread team
lwIP-2.1.2 initialized!
```

```
read W25Qxx ID is: 0x16EF
[I/sal.skt] Socket Abstraction Layer initialize success.
[I/[drv.dm9051] ] [dm9051_probe L674] device_id: 90510A46
[I/[drv.dm9051] ] [dm9051_probe L682] CHIP Revision: 01
msh />[I/[drv.dm9051] ] MAC: 00-60-6E-10-34-54
[I/[drv.dm9051] ] link up!
ka_mqtt
msh />
[I] >> The ID of the Kawaii client is: rtthread19874
[I] >> ../packages/kawaii-mqtt-latest/mqttclient/mqttclient.c:991 mqtt_connect_with_results()... mqtt connect
success...
    [I] >> -----------------------------------------------------------------------------------
    [I] >> ../packages/kawaimqtt-latest/test/test.c:20 sub_topic_handle1()...
topic: rtt-sub
message:{
    "msg": "hello"
}
    [I] >> -----------------------------------------------------------------------------------
```

6.3.4　小结

本节主要介绍 MQTT 协议的基本原理和基本概念。通过本节的学习，读者可掌握 MQTT 协议管理接口的应用。

6.4　HTTP 应用开发

超文本传输协议（Hypertext Transfer Protocol，HTTP）是一种用于在计算机之间传输超文本的应用层协议，它是万维网（World Wide Web）的核心协议，用于在客户端和服务器之间传输数据。HTTP 是一个无状态的协议，意味着每个 HTTP 请求都是独立的，服务器不会在不同请求之间保持客户端的状态信息。

本节的要求如下：

➲ 了解 HTTP 基本概念与工作原理。

➲ 掌握 HTTP 管理方式。

➲ 掌握 RT-Thread HTTP 管理接口的应用。

6.4.1　原理分析

6.4.1.1　HTTP 的工作原理

HTTP 是互联网中应用最为广泛的一种网络协议，是一种请求-响应式的协议。HTTP 数据交互流程是：一个客户端与服务器建立连接之后，客户端发送一个请求给服务器，服务器接收到请求后根据接收到的信息确定响应方式，并且给予客户端相应的响应。

浏览器网页是 HTTP 的主要应用方式，但这不表示 HTTP 只能应用于网页。实际上，只要通信双方遵循 HTTP，就可以进行数据交互，如嵌入式系统通过 HTTP 与服务器连接。HTTP 的工作原理如图 6.30 所示。

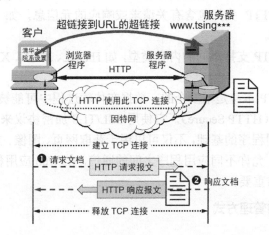

图 6.30　HTTP 的工作原理

6.4.1.2　HTTP 的基本概念

HTTP 是一种无状态协议。无状态是指协议对于事件的处理没有记忆能力，这意味着如果后续处理需要先前的信息，则必须重传先前的信息，这可能导致每次连接传输的数据量变大，但这样做的好处是服务器不需要先前的信息，应答很快。HTTP 允许传输任意类型的数据，传输的数据类型由 Content-Type 加以标记。当客户端向服务器发送请求时，只需要发送请求方式和路径即可。由于 HTTP 非常简单，使得 HTTP 服务器的程序规模小，通信速率很快。HTTP 支持 C/S（客户机/服务器）架构和 B/S（浏览器/服务器）架构。

HTTP 的基本概念：

（1）采用客户端-服务器模型。HTTP 采用客户端-服务器模型，客户端发送请求，服务器响应该请求。客户端通常是浏览器，服务器是存储和提供网页的远程计算机。

（2）采用请求-响应模式。HTTP 通信由请求和响应组成，客户端向服务器发送请求，请求包括请求方法、URL、HTTP 版本号、请求头和请求体，服务器接收到请求后，会产生响应，响应包括 HTTP 状态码、响应头和响应体。

（3）HTTP 状态码。用于指示请求的结果，常见的 HTTP 状态码包括：

⊃ 200：请求成功。

⊃ 404：请求的资源未找到。

⊃ 500：服务器发生错误。

（4）无状态。HTTP 是一种无状态协议，每个请求都是独立的，服务器不会记住之前的请求。为了处理会话状态，通常使用 Cookies 或会话标识符来跟踪客户端状态。

（5）URL（Uniform Resource Locator）。用于标识请求资源的地址，包括协议、主机、端口和路径信息。

（6）请求方法。HTTP 定义了不同的请求方法，主要包括：

⊃ GET：用于请求资源。

⊃ POST：用于向服务器提交数据。

⊃ PUT：用于更新资源。

͝⊃ DELETE：用于删除资源。

͝⊃ HEAD：用于获取资源的响应头信息。

（7）头部字段。HTTP 头部包含有关请求或响应的元信息，如 Content-Type、Content-Length、User-Agent 等。

（8）内容类型。HTTP 支持不同的内容类型，如 HTML、JSON、XML 等，由 Content-Type 确定。

（9）安全性。HTTP 本身是不安全的，数据在传输过程中可能被窃听。为了安全传输数据，通常使用 HTTPS（HTTP Secure），它使用 SSL/TLS 加密协议来保护通信。

HTTP 是 Web 应用程序的基础，不仅可以用于获取网页、图像、文档和其他多媒体资源，还可以用于 Web 服务，允许不同应用程序之间的通信。在 Web 应用程序开发中，了解 HTTP 的工作原理和用法非常重要。

6.4.1.3　HTTP 的管理方式

1）创建会话

通过下面的函数可创建会话：

```
/*********************************************************************************
 * 名称：webclient_session_create()
 * 功能：创建客户端会话结构体
 * 参数：header_sz 表示支持的头部最大长度
 * 返回：非 NULL 表示指向会话结构体的指针；NULL 表示创建会话失败
 *********************************************************************************/
struct webclient_session *webclient_session_create(size_t header_sz);
```

2）发送 GET 请求

通过下面的函数可发送 GET 请求：

```
/*********************************************************************************
 * 名称：webclient_get()
 * 功能：发送 GET 请求命令
 * 参数：session 表示指向当前会话结构体的指针；URI 表示连接的 HTTP 服务器地址
 * 返回：大于 0 的值表示 HTTP 状态码；小于 0 的值表示发送 GET 请求失败
 *********************************************************************************/
int webclient_get(struct webclient_session *session, const char *URI);
```

3）获取 Content-Length 字段数据

在发送 GET 或 POST 请求后，通过下面的函数可获取返回的 Content-Length 字段数据：

```
/*********************************************************************************
 * 名称：webclient_content_length_get()
 * 功能：获取 Content-Length 字段数据
 * 参数：session 表示指向当前会话结构体的指针
 * 返回：大于 0 的值表示 Content-Length 字段数据；小于 0 的值表示获取 Content-Length 字段数据失败
 *********************************************************************************/
int webclient_content_length_get(struct webclient_session *session);
```

4）接收数据

通过下面的函数可接收数据：

```
/***********************************************************************
 * 名称：webclient_read()
 * 功能：接收数据
 * 参数：session 表示指向当前会话结构体的指针；buffer 表示接收数据的缓存地址；size 表示允许接
收数据的最大长度
 * 返回：大于 0 的值表示成功接收到的数据长度；0 表示连接关闭；小于 0 的值表示接收数据失败
 ***********************************************************************/
int webclient_read(struct webclient_session *session, unsigned char *buffer, size_t size);
```

5）关闭会话

通过下面的函数可关闭会话：

```
/***********************************************************************
 * 名称：webclient_close()
 * 功能：关闭会话
 * 参数：session 表示指向当前会话结构体的指针
 * 返回：0 表示成功
 ***********************************************************************/
int webclient_close(struct webclient_session *session);
```

6.4.2 开发设计与实践

6.4.2.1 硬件设计

同 3.2.2.1 节。

6.4.2.2 软件设计

软件设计流程如图 6.31 所示。

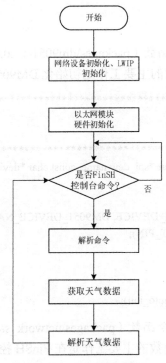

图 6.31 软件设计流程

6.4.2.3　功能设计与核心代码设计

本节以获取天气数据为例介绍 HTTP 管理接口的应用。要通过 HTTP 获取天气数据，首先需要初始化以太网模块硬件；然后使用 FinSH 控制台命令控制 HTTP 通信，需要用户自定义 FinSH 控制台命令，并将已实现的命令添加到 FinSH 控制台命令列表中，这样在启动 MobaXterm 串口终端 FinSH 控制台后就可以使用自定义的命令了。

1）主函数（zonesion/app/main.c）

主函数的主要工作是初始化以太网模块硬件，初始化完成后退出主函数。代码如下：

```c
/****************************************************************************
* 名称：main()
* 功能：初始化以太网模块硬件
*****************************************************************************/
#include <rtthread.h>

#define DBG_TAG "main"
#define DBG_LVL DBG_LOG
#include <rtdbg.h>

#include "drv_spi.h"
int dm9051_auto_init(void);
int main(void)
{
    rt_hw_spi_device_attach("spi1", "spi11", GPIOD, GPIO_PIN_3);
    //网络设备初始化、LWIP 初始化已经在系统初始化时自动完成
    dm9051_auto_init();                                         //以太网模块硬件初始化
    return 0;
}
```

2）以太网模块硬件初始化函数（packages\dm9051-v1.0.0\src\drv_dm9051_init.c）

以太网模块硬件初始化函数的主要工作是初始化 DM9051。代码如下：

```c
/****************************************************************************
* 名称：dm9051_auto_init()
* 功能：初始化以太网模块硬件
*****************************************************************************/
extern int dm9051_probe(const char *spi_dev_name, const char *device_name, int rst_pin, int int_pin);
int dm9051_auto_init(void)
{
    dm9051_probe(DM9051_SPI_DEVICE, DM9051_DEVICE_NAME, DM9051_RST_PIN,
            DM9051_INT_PIN);

    return 0;
}
//INIT_ENV_EXPORT(dm9051_auto_init);
```

3）自定义 FinSH 控制台命令函数（packages\network_samples-latest\httpclient_sample.c）

自定义 FinSH 控制台命令函数的主要工作是在 FinSH 控制台中使用命令获取天气数据。代码如下：

```
/******************************************************************************
* 名称：weather(int argc, char **argv)
* 功能：获取天气数据的命令
* 参数：argc 表示参数个数；argv 表示指向参数字符串指针的数组指针
******************************************************************************/
void weather(int argc, char **argv)
{
    rt_uint8_t *buffer = RT_NULL;
    int resp_status;
    struct webclient_session *session = RT_NULL;
    char *weather_url = RT_NULL;
    int content_length = -1, bytes_read = 0;
    int content_pos = 0;
    //为 weather_url 分配空间
    weather_url = rt_calloc(1, GET_URL_LEN_MAX);
    if (weather_url == RT_NULL)
    {
        rt_kprintf("No memory for weather_url!\n");
        goto __exit;
    }
    //拼接 GET 请求中的网址
    if(argv[1]){
      rt_snprintf(weather_url, GET_URL_LEN_MAX, GET_URI, argv[1]);
      rt_kprintf("weather_url use custom AREA_ID:%s!\n",argv[1]);
      rt_kprintf("weather_url:%s\n",weather_url);
    }
    else{
      rt_snprintf(weather_url, GET_URL_LEN_MAX, GET_URI, AREA_ID);
      rt_kprintf("weather_url use default AREA_ID:hubei wuhan(101200101)!\n");
      rt_kprintf("weather_url:%s\n",weather_url);
    }
    //创建会话
    session = webclient_session_create(GET_HEADER_BUFSZ);
    if (session == RT_NULL)
    {
        rt_kprintf("No memory for get header!\n");
        goto __exit;
    }
    //发送 GET 请求，使用默认的头部
    if ((resp_status = webclient_get(session, weather_url)) != 200)
    {
        rt_kprintf("webclient GET request failed, response(%d) error.\n", resp_status);
        goto __exit;
    }
    //分配用于存放天气数据的缓冲区
    buffer = rt_calloc(1, GET_RESP_BUFSZ);
    if (buffer == RT_NULL)
```

```
    {
        rt_kprintf("No memory for data receive buffer!\n");
        goto __exit;
    }
    content_length = webclient_content_length_get(session);
    if (content_length < 0)
    {
        //返回的数据是分块的
        do
        {
            bytes_read = webclient_read(session, buffer, GET_RESP_BUFSZ);
            if (bytes_read <= 0)
            {
                break;
            }
        }while (1);
    }
    else
    {
        do
        {
            bytes_read = webclient_read(session, buffer,
                                content_length - content_pos > GET_RESP_BUFSZ ?
                                GET_RESP_BUFSZ : content_length - content_pos);
            if (bytes_read <= 0)
            {
                break;
            }
            content_pos += bytes_read;
        }while (content_pos < content_length);
    }
    //解析天气数据
    weather_data_parse(buffer);
__exit:
    //释放网址空间
    if (weather_url != RT_NULL)
    rt_free(weather_url);
    //关闭会话
    if (session != RT_NULL)
    webclient_close(session);
    //释放缓冲区
    if (buffer != RT_NULL)
    rt_free(buffer);
}
MSH_CMD_EXPORT(weather, Get weather by webclient);
```

4）天气数据解析函数（packages\network_samples-latest\httpclient_sample.c）

天气数据解析函数的主要工作是对接收到的天气数据进行解析。代码如下：

```
/*******************************************************************************
* 名称：weather_data_parse()
* 功能：解析接收到的天气数据
* 参数：data 表示接收到的天气数据
*******************************************************************************/
void weather_data_parse(rt_uint8_t *data)
{
    cJSON *root = RT_NULL, *item = RT_NULL;
    root = cJSON_Parse((const char *)data);
    if (!root)
    {
        rt_kprintf("No memory for cJSON root!\n");
        return;
    }
    item = cJSON_GetObjectItem(root, "cityEn");              //城市名称
    rt_kprintf("\ncityEn     :%s ", item->valuestring);

    item = cJSON_GetObjectItem(root, "wea");                 //天气情况
    rt_kprintf("\nwea          :%s ", item->valuestring);

    item = cJSON_GetObjectItem(root, "tem");                 //实时温度
    rt_kprintf("\ntem          :%s ", item->valuestring);

    item = cJSON_GetObjectItem(root, "tem2");                //高温
    rt_kprintf("\ntem_range :%s", item->valuestring);

    item = cJSON_GetObjectItem(root, "tem1");                //低温
    rt_kprintf(" ~ %s", item->valuestring);

    item = cJSON_GetObjectItem(root, "humidity");            //湿度
    rt_kprintf("\nhumidity    :%s ", item->valuestring);

    item = cJSON_GetObjectItem(root, "win");                 //风向
    rt_kprintf("\nwin          :%s ", item->valuestring);

    item = cJSON_GetObjectItem(root, "win_speed");           //风力等级
    rt_kprintf("\nwin_speed :%s ", item->valuestring);

    item = cJSON_GetObjectItem(root, "win_meter");           //风速
    rt_kprintf("\nwin_meter :%s ", item->valuestring);

    item = cJSON_GetObjectItem(root, "air");                 //空气质量
    rt_kprintf("\nair          :%s ", item->valuestring);

    item = cJSON_GetObjectItem(root, "air_level");           //空气质量等级
    rt_kprintf("\nair_level :%s ", item->valuestring);
```

```
        item = cJSON_GetObjectItem(root, "update_time");          //气象台更新时间
        rt_kprintf("\nupdate_time:%s \n", item->valuestring);
        if (root != RT_NULL)
            cJSON_Delete(root);
}
```

6.4.3　开发步骤与验证

6.4.3.1　硬件部署

同 1.2.3.1 节。

6.4.3.2　工程调试

同 2.1.3.2 节。

6.4.3.3　验证效果

（1）关闭 RT-Thread Studio 软件，拔掉仿真器。按下 ZI-ARMEmbed 电源重新开机上电。

（2）在 MobaXterm 串口终端 FinSH 控制台中输入"weather"命令，可获取默认城市的天气数据（这里默认的城市是武汉），并解析天气数据。

```
 \ | /
- RT -       Thread Operating System
 / | \       4.1.0 build Nov 15 2022 16:36:31
 2006 - 2022 Copyright by RT-Thread team
lwIP-2.1.2 initialized!
read W25Qxx ID is: 0xEF16
[I/sal.skt] Socket Abstraction Layer initialize success.
[I/[drv.dm9051] ] [dm9051_probe L674] device_id: 90510A46
[I/[drv.dm9051] ] [dm9051_probe L682] CHIP Revision: 01
msh >[I/[drv.dm9051] ] MAC: 00-60-6E-10-34-54
[I/[drv.dm9051] ] link up!
msh >weather
weather_url use default AREA_ID:hubei wuhan(101200101)!
weather_url:http://v0.yiketianqi.com/api?version=v61&appid=38119558&appsecret=8k2oHWWA&cityid=1
01200101

cityEn      :wuhan
wea         :雾
tem         :14
tem_range :10 ~ 16
humidity    :81%
win         :东北风
win_speed :1 级
win_meter :4km/h
air         :61
air_level :良
update_time:09:55
msh >
```

（3）在 MobaXterm 串口终端 FinSH 控制台中输入 "weather 101010100" 命令，可获取北京的天气数据（101010100 为北京的 ID），并解析天气数据。

```
    \|/
  - RT -      Thread Operating System
   /|\     4.1.0 build Nov 15 2022 16:36:31
  2006 - 2022 Copyright by RT-Thread team
lwIP-2.1.2 initialized!
read W25Qxx ID is: 0xEF16
[I/sal.skt] Socket Abstraction Layer initialize success.
[I/[drv.dm9051] ] [dm9051_probe L674] device_id: 90510A46
[I/[drv.dm9051] ] [dm9051_probe L682] CHIP Revision: 01
msh >[I/[drv.dm9051] ] MAC: 00-60-6E-10-34-54
[I/[drv.dm9051] ] link up!
msh >weather
weather_url use default AREA_ID:hubei wuhan(101200101)!
weather_url:http://v0.yiketianqi.com/api?version=v61&appid=38119558&appsecret=8k2oHWWA&cityid=1
01200101

    cityEn       :wuhan
    wea          :雾
    tem          :14
    tem_range :10 ~ 16
    humidity     :81%
    win          :东北风
    win_speed :1 级
    win_meter :4km/h
    air          :61
    air_level :良
    update_time:09:55
msh >weather 101010100
weather_url use custom AREA_ID:101010100!
weather_url:http://v0.yiketianqi.com/api?version=v61&appid=38119558&appsecret=8k2oHWWA&cityid=1
01010100

    cityEn       :beijing
    wea          :多云
    tem          :10
    tem_range :1 ~ 14
    humidity     :34%
    win          :东北风
    win_speed :2 级
    win_meter :7km/h
    air          :27
    air_level :优
    update_time:09:50
msh >
```

（4）在 MobaXterm 串口终端 FinSH 控制台中输入"weather 101020100"命令，可获取
上海的天气数据（101020100 为上海的 ID），并解析天气数据。

```
       \|/
     - RT -        Thread Operating System
      /|\      4.1.0 build Nov 15 2022 16:36:31
     2006 - 2022 Copyright by RT-Thread team
lwIP-2.1.2 initialized!
read W25Qxx ID is: 0xEF16
[I/sal.skt] Socket Abstraction Layer initialize success.
[I/[drv.dm9051] ] [dm9051_probe L674] device_id: 90510A46
[I/[drv.dm9051] ] [dm9051_probe L682] CHIP Revision: 01
msh >[I/[drv.dm9051] ] MAC: 00-60-6E-10-34-54
[I/[drv.dm9051] ] link up!
msh >weather
weather_url use default AREA_ID:hubei wuhan(101200101)!
weather_url:http://v0.yiketianqi.com/api?version=v61&appid=38119558&appsecret=8k2oHWWA&cityid=1
01200101

    cityEn     :wuhan
    wea        :雾
    tem        :14
    tem_range :10 ~ 16
    humidity   :81%
    win        :东北风
    win_speed :1 级
    win_meter :4km/h
    air        :61
    air_level :良
    update_time:09:55
    msh >weather 101010100
    weather_url use custom AREA_ID:101010100!
    weather_url:http://v0.yiketianqi.com/api?version=v61&appid=38119558&appsecret=8k2oHWWA&cityid=1
01010100

    cityEn     :beijing
    wea        :多云
    tem        :10
    tem_range :1 ~ 14
    humidity   :34%
    win        :东北风
    win_speed :2 级
    win_meter :7km/h
    air        :27
    air_level :优
    update_time:09:50
    msh >weather 101020100
    weather_url use custom AREA_ID:101020100!
```

```
weather_url:http://v0.yiketianqi.com/api?version=v61&appid=38119558&appsecret=8k2oHWWA&cityid=1
01020100

cityEn      :shanghai
wea         :多云
tem         :15
tem_range :13 ~ 18
humidity    :48%
win         :北风
win_speed :0 级
win_meter :0km/h
air         :122
air_level :轻度污染
update_time:10:18
msh >
```

6.4.4　小结

本节主要介绍 HTTP 协议的基本原理和基本概念。通过本节的学习，读者可掌握 HTTP 协议管理接口的应用。

参考文献

[1] 杨洁. RT-Thread 设备驱动开发指南[M]. 北京：机械工业出版社，2023.

[2] 刘火良，杨森. RT-Thread 内核实现与应用开发实战指南[M]. 北京：机械工业出版社，2018.

[3] 杨洁，郭占鑫，刘康，等. RT-Thread 内核实现与应用开发实战指南[M]. 北京：机械工业出版社，2022.

[4] 邱祎. 嵌入式实时操作系统：RT-Thread 设计与实现[M]. 北京：机械工业出版社，2022.

[5] 廖建尚，王治国，郝玉胜. 嵌入式 Linux 开发技术[M]. 北京：电子工业出版社，2021.

[6] 鸟哥. 鸟哥的 Linux 私房菜基础学习篇[M]. 4 版. 北京：人民邮电出版社，2018.

[7] 宋宝华. Linux 设备驱动开发详解：基于最新的 Linux4.0 内核[M]. 北京：机械工业出版社，2015.